W9-DIW-087

# BOILER OPERATIONS

# BOILER OPERATIONS

### Billy C. Langley

Harcourt Brace Jovanovich, Publishers
Technology Publications

San Diego   New York   Chicago   Austin   Washington, D.C.
London   Sydney   Tokyo   Toronto

*024427*

# PREFACE

The proficient, efficient, and, above all, safe operation of boiler rooms requires trained technicians. *Boiler Operations* was written to help train new technicians as well as to update the knowledge and skills of present personnel.

The book's 16 chapters can be covered in one term. Each chapter addresses a separate aspect of boiler plant maintenance and operation. Unlike many of the books available, *Boiler Operations* stresses safety, devoting all of Chapter 2 to the subject and emphasizing key points of safety where appropriate in other chapters.

Since students come to this course with varying degrees of mathematical ability, Chapter 1 reviews the necessary math background. If an understanding of mathematical concepts presents no problem to the reader, the chapter may be used only for reference.

Readers interested in the administration will learn the fundamentals of boiler room management in Chapter 14. This chapter also contains information about career opportunities and possibilities for advancement.

Each chapter opens with fundamentals and progresses to more advanced theory, with specific applications following the theory. Review questions at the end of each chapter reinforce concepts and test skills.

*Boiler Operations* is a comprehensive introduction to the field; the reader who absorbs the information presented here should find solid employment and advancement opportunities.

B. C. L.

# CONTENTS

# CHAPTER 1

# SHOP MATHEMATICS

## Learning Objectives

The objectives of this chapter are to:

- Introduce the reader to the different basic mathematical functions
- Provide reasons why boiler operators need to know basic mathematical operations
- Show the reader that the rules of mathematics are true
- Provide the reader with the knowledge necessary to complete basic mathematical operations.

Mathematics is a basic tool. Some version of it is found in almost everything that we do, from the simple arithmetic of counting objects to the complicated equations encountered in computer and engineering work. Boiler operators need math for much of their work. For example, they need it to compute the center of gravity for lifting objects, to determine the efficiency of a boiler, and to figure out the amount and cost of the fuel used for boiler operation. As boiler equipment becomes more complex, mathematics achieves ever-increasing importance as an essential tool.

From the point of view of the individual, there are many incentives for learning the subject. Mathematics better equips him to do his present job. It will help him attain promotions and the corresponding pay increases. In fact, it has been found that one of the best indicators of a man's potential success is his understanding of mathematics.

This chapter begins with an overview of basic arithmetic and continues through the decimal system. An attempt is made throughout to explain why rules of mathematics are true. This is done because it is felt that the rules are easier to learn and remember if the ideas that led to their development are understood.

## NUMBER SYSTEMS AND SETS

Many of us have areas in our mathematics background that are hazy, barely understood, or troublesome. While it may seem beneath our dignity to study fundamental arithmetic, the basic concepts may be just the areas where our difficulties lie.

### Counting

Counting is such a basic and natural process that we rarely stop to think about it. The process is based on the idea of *one-to-one correspondence,* which is easily demonstrated by using the fingers. When children count on their fingers, they are placing each finger in a one-to-one correspondence with each object being counted. Having outgrown finger counting, we use numerals.

**Numerals.**    Numerals are number symbols. One of the simplest systems is the Roman numeral system, in which tally marks are used to represent the objects being counted. Roman numerals appear to be a refinement of the tally method, which is still in use today. By this method, one makes short vertical marks until a total of four is reached. When a fifth tally is counted, a diagonal mark is drawn through the first four marks. Grouping by fives in this way is reminiscent of the Roman numeral system, in which the multiples of five are represented by special symbols.

A number may have many "names." For example, the number six may be indicated by any of the following symbols: *9−3, 12÷2, 5+1,* or *2×3.* The important thing to remember is that a number is an idea. Various symbols used to indicate a number are merely different ways of expressing the same idea.

**Positive Whole Numbers.**   The numbers that are used for counting in our number system are sometimes called natural numbers. They are positive whole numbers, or to use the more precise mathematical term, positive *integers.* The Arabic numerals from 0 through 9 are called digits, and an integer may be any number of digits. For example, 5, 32, and 7,049 are all integers. The number of digits in an integer indicates its rank; that is, whether it is "in the hundreds," "in the thousands," and so forth. The idea of ranking numbers in terms of tens, hundreds, thousands, etc., is based on the *place value* concept.

## Place Value

Although a system such as the Roman numeral system is adequate for recording the results of counting, it is too cumbersome for purposes of calculation. Before arithmetic could develop as we know it today, the following two important concepts were needed as additions to the counting process:

1. The idea of 0 as a number
2. Positional notation (place value)

Positional notation is a form of coding in which the value of each digit of a number depends upon its position in relation to the other digits of the number. The convention used in our number system is that each digit has a higher place value than those to the right of it.

The place value which corresponds to a given position in a number is determined by the *base* of the numbering system. The base which is most commonly used is ten, and the system with ten as a base is called the decimal system. (Decimal is the Latin word for ten.) Any number is assumed to be a base-ten number, unless some other base is indicated. One exception to this rule occurs when the subject of an entire discussion is some base other than ten. For example, in the discussion of binary (base two) numbers later in this chapter, all numbers are assumed to be binary numbers unless some other base is indicated.

**Decimal System.**   In the United States, we use the decimal system almost exclusively. In the decimal system, each digit position in a number has ten times the value of the position adjacent to it on the right. For example, in the number 11, the 1 on the left is said to be "in the tens place," and its value is ten times as great as that of the 1 on the right. The 1 on the right is said to be "in the units place," with the understanding that the term *unit* in our system refers to the numeral 1. Thus the number 11 is actually a coded symbol which means "one ten plus one unit." Since ten plus one is eleven, the symbol 11 represents the number eleven.

The placement of the decimal point is extremely important when calculating quantities such as the amount of fuel used, the cost of fuel, the amount of steam produced per unit of fuel, or the cost of man-hours used per job or week.

Figure 1-1 shows the names of several digit positions in the decimal system. If we apply this nomenclature to the digits of the integer 235, then this number symbol means "two hundred plus three tens plus five units." This number may be expressed in mathematical symbols as follows:

$$2 \times 10 \times 10 + 3 \times 10 + 5 \times 1$$

Notice that this bears out our earlier statement: Each digit position has 10 times the value of the position adjacent to it on the right.

The integer 4,372 is a number symbol whose meaning is "four thousands plus three hundreds plus seven tens plus two units." Expressed in mathematical symbols, this number is as follows:

$$4 \times 1000 + 3 \times 100 + 7 \times 10 + 2 \times 1$$

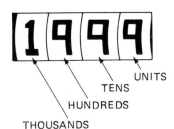

UNITS
TENS
HUNDREDS
THOUSANDS

**FIGURE 1-1.**
Names of digit positions

This presentation may be broken down further, in order to show that each digit position has 10 times the place value of the position on its right:

$$4 \times 10 \times 100 + 3 \times 10 \times 10 + 7 \times 10 \times 1 + 2 \times 1$$

The comma which appears in a number symbol such as 4,372 is used for *pointing off,* or separating the digits into groups of three beginning at the right-hand side. The first group of three digits on the right is the units group; the second group is the thousands group; the third group is the millions group, etc. Some of these groups are shown in Table 1–1.

By referring to Table 1–1, we can verify that 5,432,786 is read as follows: five million, four hundred thirty-two thousand, seven hundred eighty-six. Notice that the word *and* is not necessary when reading numbers of this kind.

**Binary System.**    The binary system is constructed in the same manner as the decimal system. However, since the base of this system is two, only two digit symbols are needed for writing numbers. These two digits are 1 and 0. In order to understand why only two digit symbols are needed in the binary system, we may make some observations about the decimal system and then generalize from these.

One of the most striking observations about number systems which utilize the concept of place value is that there is no single-digit symbol for the base. For example, in the decimal system the symbol for ten, the base, is 10. This symbol is compounded from two digit symbols, and its meaning may be interpreted as "one base plus no units." Notice the implication of this where other bases are concerned: Every system used the same symbol for the base, namely 10. Furthermore, the symbol 10 is not called "ten" except in the decimal system.

Suppose that a number system were constructed with five as the base. Then the only digit symbols needed would be 0, 1, 2, 3, and 4. No single-digit symbol for five is needed, since the symbol 10 in a base-five system

**TABLE 1–1**
Place Values and Grouping

| Billions group | Millions group | Thousands group | Units group |
|---|---|---|---|
| Hundred billions | Hundred millions | Hundred thousands | Hundreds |
| Ten billions | Ten millions | Ten thousands | Tens |
| Billions | Millions | Thousands | Units |

with place value means "one five plus no units." In general, in a number system using base *N,* the largest number for which a single-digit symbol is needed is *N* minus 1. Therefore, when the base is two the only digit symbols needed are 1 and 0.

An example of a binary number is the symbol 101. We can discover the meaning of this symbol by relating it to the decimal system. (See Figure 1–2.) Here the place value of each digit position in the binary system is two times the place value of the position adjacent to it on the right. Compare this to Figure 1–1, in which the base is ten rather than two.

Placing the digits of the number 101 in their respective blocks on Figure 1–2, we find that 101 means "one four plus no twos plus one unit." Thus 101 is the binary equivalent of decimal 5. If we wish to convert a decimal number, such as 7, to its binary equivalent, we must break it into parts which are multiples of 2. Since 7 is equal to 4 plus 2 plus 1, we say that it contains one four, one two, and one unit. Therefore, the binary symbol for decimal 7 is 111.

The most common use of the binary number system is in electronic digital computers. All data fed to a typical electronic digital computer is converted to binary form and the computer performs its calculations using binary arithmetic rather than decimal arithmetic. One of the reasons for this is that electrical and electronic equipment utilizes many switching circuits in which there are only two operating conditions. Either the circuit is *on* or it is *off,* and a two-digit number system is ideally suited for symbolizing such a situation.

Details concerning binary arithmetic are beyond the scope of this text, but are available in other books.

## Sets

Any serious study of mathematics involves reviewing more than one text and discovering that there is more than one way to approach each new

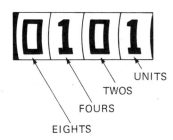

**FIGURE 1-2.**

Digit positions in the binary system

topic. In the following paragraphs, a very brief introduction to some of the set theory in modern math is presented. This is intended to help students make the transition from traditional methods to newer methods.

**Definitions and Symbols.**   The word *set* implies a collection or grouping of similar objects or symbols. The objects in a set have at least one characteristic in common, such as similarity of appearance or purpose. A set of tools would be an example of a group of objects not necessarily similar in appearance but similar in purpose. The objects or symbols in a set are called members or *elements* of the set.

The elements of a mathematical set are usually symbols, such as numerals, lines, or points. For example, the integers greater than 0 and less than 5 form a set, as follows:

$$[1, 2, 3, 4]$$

Notice that braces are used to denote sets. Brackets are used when the elements of a set are not too numerous.

Since the elements of the set [2, 4, 6] are the same as the elements of [4, 2, 6], these two sets are said to be equal. Equality between sets has nothing to do with the order in which the elements are arranged. Furthermore, it is not necessary to repeat elements. The elements of [2, 2, 3, 4] are simply 2, 3, and 4. Therefore, the sets [2, 3, 4] and [2, 2, 3, 4] are equal.

**Subsets.**   Since it is inconvenient to enumerate all of the elements of a set each time the set is mentioned, sets are often designated by a letter. For example, we could let $S$ represent the set of all integers greater than 0 and less than 10. In symbols, this relationship could be stated as follows:

$$S = [1, 2, 3, 4, 5, 6, 7, 8, 9]$$

Now suppose that we have another set, $T$, which comprised all positive even integers less than 10. This set is then defined as:

$$T = [2, 4, 6, 8]$$

Notice that every element of T is also an element of S. This establishes the subset relationship; $T$ is said to be a subset of $S$.

**Positive Integers.**   The most fundamental set of numbers is the set of positive integers. This set comprises the counting numbers (natural numbers) and includes, as subsets, all of the sets of numbers which we have discussed.

The set of natural numbers has an outstanding characteristic: it is infinite. This means that the successive elements of the set continue to increase in size without limit, each number being larger by 1 than the number preceding it. Therefore there is no largest number; any number that we might choose as larger than all the others could be increased to a larger number by simply adding 1 to it.

One way to represent the set of natural numbers symbolically would be as follows:

$$[1, 2, 3, 4, 5, 6, \ldots]$$

The three dots, called *ellipsis points,* indicate that the pattern established by the numbers shown continues without limit. In other words, the next number in the set is understood to be 7, the next after that is 8, and so forth.

**Points and Lines.**    In addition to the many sets which can be formed with number symbols, we frequently find it necessary in mathematics to work with sets composed of points or lines.

A point, like a number, is an idea rather than a tangible object. The mark that is made on a piece of paper is merely a symbol representing the point. In strict mathematical terms, a point has no dimensions (physical size) at all. Thus a pencil dot is only a rough picture of a point, and is used to indicate the location of the point.

Suppose a large number of points are placed side by side to form a "string." If we picture this arrangement by drawing dots on paper, we have a "dotted line." If more dots were placed between the dots already in the string, with the number of dots increasing until we could not see between them, we would have a rough picture of a line. Once again, it is important to emphasize that the picture is only a symbol which represents an ideal line. The ideal line would have length but no width or thickness.

The foregoing discussion leads to the conclusion that a line is actually a set of points. The number of elements in the set is infinite, since the line extends in both directions without limit.

The idea of arranging points together to form a line may be extended to the formation of planes (flat surfaces). A mathematical plane is determined by the three points which do not lie on the same line. It is also determined by two intersecting lines.

*Line Segments and Rays.*    When we draw a "line," label its end points A and B, and call it "line AB," we really mean *line segment* AB. A line segment is a subset of the set of points comprising a line.

When a line is considered to have a starting point but no stopping point (that is, it extends without limit in one direction), it is called a *ray.* A ray is not a line segment, because it does not terminate at both ends. It may be appropriate to refer to a ray as a "half-line."

**The Number Line.**    As in the case of a line segment, a ray is a subset of the set of points comprising a line. All three—lines, line segments, and rays—are subsets of the set of points comprising a plane.

Among the many devices used for representing a set of numbers, one of the most useful is the number line. To illustrate the construction of a number line, let us place the elements of the set of natural numbers in one-to-one correspondence with points on a line. Since the natural numbers are equally spaced, we select points such that the distances between them are equal. The starting point is labeled 0, the next point is labeled 1, the next 2, and so forth, using the natural numbers in normal counting order. (See Figure 1–3.) Such an arrangement is often referred to as a scale, a familiar example being the scale on a thermometer.

So far in our discussion, we have not mentioned any numbers other than integers. The number line is an ideal device for picturing the relationship between integers and other numbers such as fractions and decimals. It is clear that many points, other than those representing integers, exist on the number line. Examples are the points representing the numbers 1/2 (located halfway between 0 and 1) and 2.5 (located halfway between 2 and 3).

An interesting question arises concerning the in-between points on the number line: How many points (numbers) exist between any two integers? To answer this question, suppose that we first locate the point halfway between 0 and 1, which corresponds to the number 1/2. Then let us locate the point halfway between 0 and 1/2, which corresponds to the number 1/4. The result of the next halving operation would be 1/8, the next 1/16, etc. If we need more space to continue our halving operations on the number line, we can enlarge our "picture" and then continue.

It soon becomes apparent that the halving process could continue indefinitely; that is, without limit. In othe words, the number of points between 0 and 1 is infinite. The same is true of any other interval on the num-

**FIGURE 1-3.**
A number line

ber line. Thus, between any two integers there is an infinite set of numbers other than integers. If this seems physically impossible, considering that even the sharpest pencil point is some width, remember that we are working with ideal points, which have no physical dimensions whatsoever.

Although it is beyond the scope of this book to discuss such topics as orders of infinity, it is interesting to note that the set of integers contains many subsets which are themselves infinite. Not only are the many subsets of numbers other than integers infinite, but so are such subsets as the set of all odd integers and the set of all even integers. By intuition we see that these two subsets are infinite, as follows. If we select a particular odd or even integer which we think is the largest possible, a larger one can be formed immediately by merely adding 2.

Perhaps the most practical use of the number line is in explaining the meaning of negative numbers. Negative numbers are discussed later in this chapter.

## POSITIVE NUMBERS

The purpose of this section is to review the methods of combining integers. We have already used one combination process in our discussion of counting. We will extend the idea of counting, which is nothing more than simple addition, to develop a systematic method for adding numbers of any size. We will also learn the meaning of subtraction, multiplication, and division.

### Addition and Subtraction

In the following discussion, it is assumed that the reader knows the basic addition and subtraction tables, which present such facts as: $2 + 3 = 5$, $9 + 8 = 17$, $8 - 3 = 5$, and so forth.

The operation of addition is indicated by a plus sign $(+)$ as in $8 + 4 = 12$. The numbers 8 and 4 are *addends* and the answer (12) is their *sum*. The operation of subtraction is indicated by a minus sign $(-)$ as in $9 - 3 = 6$. The number 9 is the *minuend,* 3 is the *subtrahend,* and the answer (6) is their *difference.*

Addition is used in boiler operation to find totals such as the number of hours of boiler operation for a given period of time, the amount of fuel used for a given period of time, or the number of hours an employee spends on a given project. Subtraction is used to determine amounts such as the amount of fuel left in storage, or the amount of water treatment solution left.

Example

If a fuel oil tank has a capacity of 5,000 gallons and it is determined that there are 2,160 gallons left in the tank, how many gallons are needed to completely refill the tank?

$$\begin{aligned} \text{Gallons needed} &= \text{Gallons capacity} - \text{gallons present} \\ &= 5000 - 2160 \\ &= 2840 \text{ gallons} \end{aligned}$$

Example

An employee worked 10 hours Monday and Tuesday and 8 hours per day the remaining three days of the week. How many hours has the employee worked this week? Assuming that he receives overtime for all hours over 40 worked per week, how many hours does he have in overtime?

$$\begin{aligned} \text{Total hours} &= \text{Hours Monday} + \text{hours Tuesday} + \text{hours Wednesday} \\ &\quad + \text{hours Thursday} + \text{hours Friday} \\ &= 10 + 10 + 8 + 8 + 8 \\ &= 20 + 24 \\ &= 44 \end{aligned}$$

$$\begin{aligned} \text{Overtime} &= \text{Total hours} - 40 \\ &= 44 - 4 \\ &= 4 \text{ hours} \end{aligned}$$

**Regrouping.**    Addition may be performed with the addends arranged horizontally, if they were small enough and not too numerous. However, the most common method of arranging the addends is to place them in vertical columns. In this arrangement, the units digits of all the addends are aligned vertically, as are the tens digits, the hundreds digits, etc. The following example shows three addends arranged properly for addition:

$$\begin{array}{r} 357 \\ 1,845 \\ \underline{22} \end{array}$$

It is customary to draw a line below the last addend, placing the answer below this line. Subtraction problems are arranged in columns in the same manner as for addition, with a line at the bottom and the answer below this line.

***Carry and Borrow.*** Problems involving several addends, with two or more digits each, usually produce sums in one or more of the columns which are greater than 9. For example, suppose that we perform the following addition:

$$
\begin{array}{r}
357 \\
845 \\
\underline{22} \\
1,224
\end{array}
$$

The answer is found by a process called *carrying*. In this process, extra digits generated when a column exceeds 9 are carried to the next column to the left, and treated as addends in that column. Carrying may be explained by grouping the original addends. For example, 357 actually means 3 hundreds plus 5 tens plus 7 units. Rewriting the problem with each addend grouped in terms of units, tens, etc., give us the following:

$$
\begin{array}{r}
300 + 50 + 7 \\
800 + 40 + 5 \\
20 + 2 \\
\hline
1,100 + 110 + 14
\end{array}
$$

The extra digit in the units column of the answer represents 1 ten. We regroup the columns of the answer so that the units column has no digits representing tens, the tens column has no digits representing hundreds, etc., as follows:

$$
\begin{aligned}
1,100 + 110 + 14 &= 1,100 + 110 + 10 + 4 \\
&= 1,100 + 120 + 4 \\
&= 1,100 + 100 + 20 + 4 \\
&= 1,200 + 20 + 4 \\
&= 1,000 + 200 + 20 + 4 \\
&= 1,224
\end{aligned}
$$

When we carry the 10 from the expression $10 + 4$ to the tens column and place it with the 110 to make 120, the result is the same as if we had added 1 to the digits 5, 4, and 2 in the tens column of the original problem. Therefore, the thought process in addition is as follows: Add the 7, 5, and 2 in the units column, getting a sum of 14. Write down the 4 in the units column of the answer and carry the 1 to the tens column. Mentally add the 1 along with the other digits in the tens column, getting a sum of 12. Write down the 2 in the tens column of the answer and carry the 1 to the hundreds column. Mentally add the 1 along with the other digits in the hundreds column, getting a sum of 12. Write down the 2 in the hundreds column of the answer and carry the 1 to the thousands column. If there were other digits in the thousands column to which the 1 could be added, the process would continue as before. Since there are no digits in the thousands column of the original problem, this final 1 is not added to anything, but is simply written in the thousands place in the answer.

The borrow process is the opposite of carrying and is used in subtraction. Borrowing is not necessary in such problems as $46 - 5$ and $58 - 53$. In the first problem, the thought process may be "5 from 6 is 1 and bring down the 4 to get the difference, 41." In the second problem, the thought process is "3 from 8 is 5" and "5 from 5 is 0," and the answer is 5. More explicitly, the subtraction process in these examples is as follows:

$$
\begin{array}{r}
40 + 6 \\
- \phantom{0}5 \\
\hline
40 + 1 = 41
\end{array}
\qquad
\begin{array}{r}
50 + 8 \\
50 + 3 \\
\hline
0 + 5 = 5
\end{array}
$$

This illustrates that we are subtracting units from units and tens from tens.

Now consider the following problem where borrowing is involved:

$$
\begin{array}{r}
43 \\
-8 \\
\hline
\end{array}
$$

If the student uses the borrowing method, he may think "8 from 13 is 5 and bring down 3 to get the difference, 35." In this case what was actually done is as follows:

$$
\begin{array}{r}
30 + 13 \\
- \phantom{0}8 \\
\hline
30 + \phantom{0}5 = 35
\end{array}
$$

A ten has been borrowed from the tens columns and combined with the 3 in the units column to make a number large enough for subtraction of the

8. Notice that borrowing to increase the value of the digit in the units column reduces the value of the digit in the tens column by 1.

Sometimes it is necessary to borrow in more than one column. For example, suppose that we wish to subtract 2,345 from 5,234. Grouping the minuend and the subtrahend in units, tens, hundreds, etc., we have the following:

$$5{,}000 + 200 + 30 + 4$$
$$2{,}000 + 300 + 40 + 5$$

Borrowing a ten from the 30 in the tens column, we regroup as follows:

$$5{,}000 + 200 + 20 + 14$$
$$2{,}000 + 300 + 40 + \phantom{0}5$$

The units column is now ready for subtraction. By borrowing from the hundreds column, we can regroup so that subtraction is possible in the tens column as follows:

$$5{,}000 + 100 + 120 + 14$$
$$2{,}000 + 300 + \phantom{0}40 + \phantom{0}5$$

In the final regrouping, we borrow from the thousands column to make subtraction possible in the hundreds column, with the following result:

$$
\begin{array}{r}
4{,}000 + 1{,}100 + 120 + 14 \\
2{,}000 + \phantom{1{,}}300 + \phantom{1}40 + \phantom{1}5 \\
\hline
2{,}000 + \phantom{1{,}}800 + \phantom{1}80 + \phantom{1}9 = 2{,}889
\end{array}
$$

In actual practice, the borrowing and regrouping are done mentally. The numbers are written in the normal manner, as follows:

$$
\begin{array}{r}
5{,}234 \\
-2{,}345 \\
\hline
2{,}889
\end{array}
$$

The following thought process is used: Borrow from the tens column making the 4 become 14. Subtracting in the units column, 5 from 14 is 9. In the tens column, we now have a 2 in the minuend as a result of the first borrowing operation. Some students find it helpful at first to cancel any digits that are reduced as a result of borrowing, jotting down the digit of next lower value just above the cancelled digit. This has been done in the following example:

$$\begin{array}{r} \cancel{4}\cancel{12} \\ 3,234 \\ -2,345 \\ \hline 2,889 \end{array}$$

After cancelling the 3, we proceed with the subtraction, one column at a time. We borrow from the hundreds column to change the 2 that we now have in the tens column into 12. Subtracting in the tens column, 4 from 12 is 8. Proceeding in this same way for the hundreds column, 3 from 11 is 8. Finally, in the thousands column, 2 from 4 is 2.

***Denominate Numbers.*** Numbers that have a unit of measure associated with them, such as yard, kilowatt, pound, pint, etc., are called denominate numbers. The word *denominate* means the numbers have been given a name; they are not just abstract symbols. To add denominate numbers, add all units of the same kind. Simplify the results, if possible. The following example illustrates the addition of 6 feet 8 inches to 4 feet 5 inches:

$$\begin{array}{r} 6 \text{ ft} \quad 8 \text{ in} \\ 4 \text{ ft} \quad 5 \text{ in} \\ \hline 10 \text{ ft} \; 13 \text{ in} \end{array}$$

Since 13 in. is the equilavent of 1 ft 1 in., we regroup the answer as 11 ft 1 in.

A similar problem would be to add 20 degrees 44 minutes 6 seconds to 13 degrees 22 minutes 5 seconds. This is illustrated as follows:

$$\begin{array}{r} 20 \text{ deg} \; 44 \text{ min} \; 6 \text{ sec} \\ 13 \text{ deg} \; 22 \text{ min} \; 5 \text{ sec} \\ \hline 33 \text{ deg} \; 66 \text{ min} \; 11 \text{ sec} \end{array}$$

This answer is regrouped as 34 deg 6 min 11 sec.

Numbers must be expressed in units of the same kind, in order to be combined. For instance, the sum of 6 kilowatts plus 11 watts is not 7 kilowatts nor is it 7 watts. The sum can only be indicated. The operation cannot be performed unless some method is used to write these numbers in units of the same value.

Subtraction of denominate numbers also involves the regrouping idea. If we wish to subtract 16 deg 8 min 2 sec from 28 deg 4 min 3 sec, for example, we would have the following arrangement:

$$28 \text{ deg } 4 \text{ min } 3 \text{ sec}$$
$$-16 \text{ deg } 8 \text{ min } 2 \text{ sec}$$

In order to subtract 8 min from 4 min we regroup as follows:

$$27 \text{ deg } 64 \text{ min } 3 \text{ sec}$$
$$-16 \text{ deg } \phantom{6}8 \text{ min } 2 \text{ sec}$$
$$\overline{11 \text{ deg } 56 \text{ min } 1 \text{ sec}}$$

*Mental Calculation.*    Mental regrouping can be used to avoid the necessity of writing down some of the steps, or rewriting in columns, when groups of one-digit or two-digit numbers are to be added or subtracted.

One of the most common devices for rapid addition is recognition of groups of digits whose sum is 10. For example, in the following problem two "ten groups" have been marked with braces:

$$
\begin{array}{l}
7 \\[4pt]
\left.\begin{array}{l} 6 \\ 4 \end{array}\right\} 10 \\[10pt]
5 \\[4pt]
\left.\begin{array}{l} 1 \\ 9 \end{array}\right\} 10
\end{array}
$$

To add this column as grouped, you would say to yourself, "7, 17, 22, 32." The thought should be just successive totals as shown above and not such cumbersome steps as "7 + 10, 17 + 2, 22 + 10, 32."

When successive digits appear in a column and their sum is less than 10, it is often convenient to think of them, too, as a sum rather than separately. Thus, when adding a column in which the sum of two successive digits is 10 or less, group them as follows:

$$
\left.\begin{array}{l} 3 \\ 1 \\ 1 \end{array}\right\} 5
$$

$$\left.\begin{array}{c}8\\1\end{array}\right)9$$

$$\left.\begin{array}{c}4\\6\end{array}\right)10$$

The thought process here might be, as shown by the grouping, "5, 14, 24."

**Subtraction.**   In an example such as $73 - 46$, the conventional approach is to place 46 under 73 and subtract units from units and tens from tens, and write only the difference without the intermediate steps. To do this, the best method is to begin at the left. Thus, in the example $73 - 46$, we take 40 from 73 and then take 6 from the result. This is done mentally, however, and the thought process would be "73, 33, 27." In the example $64 - 39$ the thought is "34, 25."

## Multiplication and Division

Multiplication and division are used in boiler operation for determining quantities such as the number of hours a given boiler will operate with a given amount of fuel or chemical, or the cost of man-hours per job.

Example

A boiler uses 1/2 gallon of a certain water treatment chemical per hour of operation. The water treatment is delivered in 55-gallon drums. The boiler has operated 17 hours since the new drum of water treatment was installed. How many more hours can the boiler operate before the water treatment drum is empty?

| Number of hours of operation per drum | = Gallons per drum × no. of hours per gallon |
| | = $55 \times 2$ |
| | = 110 hours per drum |
| Number of hours of operation left | = Total hours per drum − hours of operation |
| | = $110 - 17$ |
| | = 93 hours left |

Multiplication may be indicated by a multiplication sign (×) between two numbers, a dot between two numbers, or parentheses around one or both of the numbers to be multiplied. The following examples illustrate these methods:

$$6 \times 8 = 48$$
$$6 \cdot 8 = 48$$
$$6(8) = 48$$
$$(6)(8) = 48$$

Notice that when the dot is used to indicate multiplication, it is centered vertically on the line in order to distinguish it from a decimal point or a period, as shown in the second example above. Notice also that when parentheses are used to indicate multiplication, the numbers to be multiplied are spaced closer together than they are when the dot or × is used.

In each of the four examples just given, 6 is the *multiplier* and 8 is the *multiplicand.* Both the 6 and the 8 are *factors,* and modern texts refer to them this way. The answer in a multiplication problem is the *product.* In the examples just given, the product is 48.

Division is usually indicated by a division sign (÷) or by placing one number over another number with a line drawn between the numbers, as in the following examples:

$$8 \div 4 = 2$$
$$8/4 = 2$$

The number 8 is the *dividend,* 4 is the *divisor,* and 2 is the *quotient.*

**Multiplication Methods.**   The multiplication of whole numbers may be thought of as a short process of adding equal numbers. For example, 6(5) and 6 × 5 are read as six 5s. Of course we could write 5 six times and add, but if we learn that the result is 30 we can save time. Although the concept of adding equal numbers is quite adequate in explaining multiplication of whole numbers, it is only a special case of a more general definition, which will be explained later in the section covering multiplication of fractions.

*Grouping.*   Let us examine the process involved in multiplying 6 times 27 to get the product 162. First, we arrange the factors in the following manner:

$$
\begin{array}{r}
27 \\
\times 6 \\
\hline
162
\end{array}
$$

The thought process is as follows:

1. Six times 7 is 42. Write down the 2 and carry the 4.
2. Six times 2 is 12. Add the 4 that was carried over from step 1 and write the result, 16, beside the 2 that was written in step 1.
3. The final answer is 162.

Table 1-2 shows that the factors were grouped in units, tens, etc. The multiplication was done in three steps:

1. Six times 7 units is 42 units (or 4 tens and 2 units)
2. Six times 2 tens is 12 tens (or 1 hundred and 2 tens).
3. Then the tens were added and the product was written as 162.

In preparing numbers for multiplication as in Table 1-2, it is important to place the digits of the factors in the proper columns; that is, units must be placed in the units column, tens in the tens column, and hundreds in the hundreds column. Notice that it is not necessary to write the zero in the case of 12 tens (120) since the 1 and 2 are written in the proper columns. In

**TABLE 1-2**
Multiplying by a One-digit Number

| | Hundreds | Tens | Units |
|---|---|---|---|
| 6(27) = 162 | | 2 | 7 |
| | | | 6 |
| | | 4 | 2 |
| | 1 | 2 | |
| | 1 | 6 | 2 |

practice, the addition is done mentally, and just the product is written without the intervening steps.

Multiplying a number with more than two digits by a one-digit number, as shown in Table 1–3, involves no new ideas. Three times 6 units is 18 units (1 ten and 8 units), 3 times 0 tens is 0, and 3 times 4 hundreds is 12 hundreds (1 thousand and 2 hundreds). Notice that it is not necessary to write the 0 resulting from the step "3 times 0 tens is 0." The two terminal 0s of the number 1,200 are also omitted, since the 1 and 2 are placed in their correct columns by the position of the 4.

*Partial Products.* In the example 6(8) = 48, the multiplying can be done another way to get the correct product, as follows:

$$6(3 + 5) = 6 \times 3 + 6 \times 5$$

That is, we can break 8 into 3 and 5, multiply each of these by the other factor, and add the partial products. This idea is employed in multiplying by a two-digit number. Consider the following example:

$$\begin{array}{r} 43 \\ \times 27 \\ \hline 1,161 \end{array}$$

Breaking the 27 into 20 + 7, we have 7 units times 43, plus 2 tens times 43, as follows:

$$43(20 + 7) = (43)(20) + (43)(7)$$

**TABLE 1–3**
Multiplying a Three-digit Number by a One-digit Number

| | *Thousands* | *Hundreds* | *Tens* | *Units* |
|---|---|---|---|---|
| 3(406) = 1,218 | | 4 | 0 | 6 |
| | | | | 3 |
| | | | 1 | 8 |
| | 1 | 2 | | |
| | 1 | 2 | 1 | 8 |

Since 7 units times 43 is 301 units, and 2 tens times 43 is 86 tens, we have the following:

$$
\begin{array}{r}
43 \\
\times 27 \\
\hline
301 \\
86 \\
\hline
1{,}161
\end{array}
$$

$301 = 3$ hundreds, 0 tens, 1 unit

$86 = 8$ hundreds, 6 tens

As long as the partial products are written in the correct columns, we can multiply beginning from either the left or the right of the multiplier. Thus, multiplying from the left, we have:

$$
\begin{array}{r}
43 \\
\times 27 \\
\hline
86 \\
301 \\
\hline
1{,}161
\end{array}
$$

Multiplication by a number having more places involves no new ideas.

***End Zeros.*** The placement of partial products must be kept in mind when multiplying problems involving end zeros, as in the following example:

$$
\begin{array}{r}
27 \\
\times 40 \\
\hline
1{,}080
\end{array}
$$

We have 0 units times 27 plus 4 tens times 27, as follows:

$$
\begin{array}{r}
27 \\
\times 40 \\
\hline
0 \\
108 \\
\hline
1{,}080
\end{array}
$$

The zero in the units place plays an important part in the reading of the final product. End zeros are often called the *place holders* since their only function in the problem is to hold the digit positions which they occupy, thus helping to place the other digits in the problem correctly.

The end zero in the foregoing problem can be accounted for very nicely, and at the same time the other digits can be placed correctly by using a shortcut. This consists of offsetting the 40 one place to the right and then simply bringing down the 0, without using it as a multiplier at all. The problem would appear as follows:

$$
\begin{array}{r}
27 \\
\times\,40 \\
\hline
1{,}080
\end{array}
$$

If the problem involves a multiplier with more than one 0, the multiplier is offset as many places to the right as there are end 0s. For example, consider the following multiplication in which the multiplier, 300, has two end 0s:

$$
\begin{array}{r}
220 \\
\times\,300 \\
\hline
66{,}000
\end{array}
$$

Notice that there are as many place-holding zeros at the end of the product as there are place-holding zeros in the multiplier and the multiplicand combined.

*Placement of Decimal Points.*    In any whole number in the decimal system, there is a terminating mark, called a decimal point, at the right-hand end of the number. Although the decimal point is seldom shown except in numbers involving decimal fractions (covered later in this chapter), its location must be known. The placement of the decimal point is automatically taken care of when the end 0s are correctly placed.

**Division Methods.**    Just as multiplication can be considered as repeated addition, division can be considered as repeated subtraction. For example, if we wish to divide 12 by 4, we subtract 4 from 12 in successive steps and tally the number of times that the subtraction is performed, as follows:

$$
\begin{array}{r}
12 \\
4* \\
\hline
8 \\
4* \\
\hline
4 \\
4* \\
\hline
0
\end{array}
$$

As indicated by the asterisks used as tally marks, 4 has been subtracted 3 times. This result is sometimes described by saying that "4 is contained in 12 three times."

Since successive subtraction is too cumbersome for rapid, concise calculation, methods which treat division as the inverse of multiplication tables should lead us to an answer for a problem such as $12 \div 4$ immediately, since we know that $3 \times 4$ is 12. However, a problem such as $84 \div 4$ is not so easy to solve by direct reference to the multiplication table.

One way to divide 84 by 4 is to note that 84 is the same as 80 plus 4. Thus $84 \div 4$ is the same as $80 \div 4$ plus $4 \div 4$. In symbols, this can be indicated as follows:

$$4\overline{)80 + 4}^{\,20 + 1}$$

(When this type of division symbol is used, the quotient is written above the vinculum, as shown.) Thus, 84 divided by 4 is 21.

From the foregoing example, it can be seen that regrouping is useful in division as well as in multiplication. However, the mechanical procedure used in division does not include writing down the regrouped form of the dividend. After becoming familiar with the process, we find that the division can be performed directly, one digit at a time, with regrouping taking place mentally. The following example illustrates this:

$$
\begin{array}{r}
14 \\
4\overline{)56} \\
4 \\
\hline
16 \\
16 \\
\hline
\end{array}
$$

The thought process is as follows: "4 is contained in 5 once" (write 1 in the tens place over the 5); "one times 4 is 4" (write 4 in the tens place under the 5, take the difference, and bring down 6); and "4 is contained in 16 four times" (write down 4 in the units place over the 6). After a little practice, many people can do the work shown under the dividend mentally and write only the quotient, if the divisor has only one digit.

The divisor is sometimes too large to be contained in the first digit of the dividend. The following example illustrates a problem of this kind:

$$
\begin{array}{r}
36 \\
7\overline{\smash{\big)}252} \\
21 \\
\hline
42 \\
42 \\
\hline
\end{array}
$$

Since 2 is not large enough to contain 7, we divide 7 into the number formed by the first two digits, 25. Seven is contained 3 times in 25; we write 3 above the 5 of the dividend. Multiplying, 3 times 7 is 21; we write 21 below the first two digits of the dividend. Subtracting, 25 minus 21 is 4; we write down the 4 and bring down the 2 in the units place of the dividend. We have now formed a new dividend, 42. Seven is contained 6 times in 42; we write 6 above the 2 of the dividend. Multiplying as before, 6 times 7 is 42; we write this product below the dividend 42. Subtracting, we have nothing left and the division is complete.

*Estimation.*    When there are two or more digits in the divisor, it is not always easy to determine the first digit of the quotient. An estimate must be made, and the resulting trial quotient may be too large or too small. For example, if 1,862 is to be divided by 38, we might estimate that 38 is contained 5 times in 186 and the first digit of our trial divisor would be 5. However, multiplication reveals that the product of 5 and 38 is larger than 186. Thus we would change the 5 in our quotient to 4, and the problem would then appear as follows:

$$
\begin{array}{r}
49 \\
38\overline{\smash{\big)}1862} \\
152 \\
\hline
342 \\
342 \\
\hline
\end{array}
$$

On the other hand, suppose that we had estimated that 38 is contained in 186 only 3 times. We would then have the following:

$$
\begin{array}{r}
3 \\
38\overline{\smash{\big)}1862} \\
114 \\
\hline
72 \\
\hline
\end{array}
$$

Now, before we make any further moves in the division process, it should be obvious that something is wrong. If our new dividend is large enough to contain the divisor before bringing down a digit from the original dividend, then the trial quotient should have been larger. In other words, our estimate is too small.

Proficiency in estimating trial quotients is gained through practice and familiarity with number combinations. For example, after a little experience we realize that a close estimate can be made in the foregoing problem by thinking of 38 as "almost 40." It is easy to see that 40 is contained 4 times in 186, since 4 times 40 is 160. Also, since 5 times 40 is 200, we are reasonably certain that 5 is too large for our trial divisor.

***Uneven Division.***    In some division problems, such as $7 \div 3$, there is no other whole number that, when multiplied by the divisor, will give the dividend. We use the distributive idea to show how division is done in such a case. For example, $7 \div 3$ could be written as follows:

$$\frac{(6 + 1)}{3} = \frac{6}{3} + \frac{1}{3} = 2\frac{1}{3}$$

Thus, we see that the quotient also carries one unit that is to be divided by 3. It should be clear by now that

$$3\overline{)37} = 3\overline{)30 + 7}$$

and that this can be further reduced as follows:

$$\frac{30}{3} + \frac{6}{3} + \frac{1}{3} = 10 + 2 + \frac{1}{3} = 12\frac{1}{3}$$

In elementary arithmetic, the part of the dividend that cannot be divided evenly by the divisor is often called a *remainder* and is placed next to the quotient with the prefix *R*. Thus, in the foregoing example where the quotient was 12 and $1 \div 3$, the quotient could be written 12 *R* 1. This method of indicating uneven division is useful in the following example.

Suppose that $13 is available for the purchase of spare parts, and the parts needed cost $3 each. Four parts can be bought with the available money, and $1 will be left over. Since it is not possible to buy $1 \div 3$ of a part, expressing the result as 4 *R* 1 gives a more meaningful answer than 4 1/3.

***Placement of Decimal Points.***    In division, as in multiplication, the placement of the decimal point is important. Determining the location of the decimal point and the number of places in the quotient can be relatively simple if the work is kept in the proper columns. For example, notice the vertical alignment in the following problem:

$$
\begin{array}{r}
311 \\
31\overline{)9{,}641} \\
9\ 3 \\
\hline
34 \\
31 \\
\hline
31 \\
31 \\
\hline
\end{array}
$$

We notice that the first two places in the dividend are used to obtain the first place in the quotient. Since 3 is in the hundreds column, there are two more places in the quotient (tens place and the units place). The decimal point in the quotient is understood to be directly above the position of the decimal point in the dividend. In the example shown above, the decimal point is not shown but is understood to be immediately after the second 1.

*Checking Accuracy.* The accuracy of the division of numbers can be checked by multiplying the quotient by the divisor and adding the remainder, if any. The result should equal the dividend. Consider the following example:

$$
\begin{array}{r}
5203 \\
42\overline{)218541} \\
210 \\
\hline
85 \\
84 \\
\hline
141 \\
126 \\
\hline
15
\end{array}
\qquad
\begin{array}{rr}
\text{check:} & 5203 \\
& \times\ 42 \\
\hline
& 10406 \\
& 20812 \\
\hline
& 218526 \\
& +15 \\
\hline
& 218541
\end{array}
$$

**Denominate Numbers.** We have learned that denominate numbers are not difficult to add and subtract, provided that units, tens, hundreds, etc., are retained in their respective columns. Mulitplication and division of denominate numbers may also be performed with comparative ease by using the experience gained in addition and subtraction.

*Multiplication.* In multiplying denominate numbers by integers, no new ideas are needed. If in the problem 3(5 yd 2 ft 6 in) we remember that we can multiply each part separately to get the correct product, as in the example, $6(8) = 6(3) + 6(5)$, we can easily find the product, as follows:

$$
\begin{array}{r}
5 \text{ yd } 2 \text{ ft } \ 6 \text{ in} \\
\times \ 3 \\
\hline
15 \text{ yd } 6 \text{ ft } 18 \text{ in}
\end{array}
$$

Simplifying, this is:

$$17 \text{ yd } 1 \text{ ft } 6 \text{ in}$$

When one denominate number is multiplied by another, a question arises concerning the products of the units of measurement. The product of one unit times another of the same kind is one square unit. For example, 1 ft times 1 ft is 1 square foot, abbreviated sq ft; 2 in. times 3 in. is 6 sq in.; etc. If it becomes necessary to multiply such numbers as 2 yd 1 ft times 6 yd 2 ft, the foot units may be converted to fractions of a yard, as follows:

$$(2 \text{ yd } 1 \text{ ft})(6 \text{ yd } 2 \text{ ft}) = (2\tfrac{1}{3} \text{ yd})(6\tfrac{2}{3} \text{ yd})$$

In order to complete the multiplication, a knowledge of fractions is needed. Fractions are discussed later in this chapter.

*Division.*   The division of denominate numbers requires division of the highest digits first; and if there is a remainder, conversion to the next lower unit; and repeated division until all units have been divided.

In the example (24 gal 1 qt 1 pt) $\div$ 5, we perform the following steps:

*Step 1*
$$
\begin{array}{r}
4 \text{ gal} \\
5\overline{\smash{)}24 \text{ gal}} \\
20 \\
\hline
4 \text{ gal (left over)}
\end{array}
$$

*Step 2*   Convert the 4 gal left over to the 16 qt and add it to the 1 qt

*Step 3*
$$
\begin{array}{r}
3 \text{ qt} \\
5\overline{\smash{)}17 \text{ qt}} \\
15 \\
\hline
2 \text{ qt (left over)}
\end{array}
$$

*Step 4*   Convert the 2 qt left over to 4 pt and add to the 1 pt

*Step 5*          $\dfrac{1 \text{ pt}}{5\overline{\smash{\big)}5 \text{ pt}}}$

Therefore, 24 gal 1 qt 1 pt divided by 5 is 4 gal 3 qt 1 pt.

**Order of Operations.**   When a series of operations involving addition, subtraction, multiplication, or division is indicated, the order in which the operations are performed is important only if division is involved or if the operations are mixed. A series of individual additions, subtractions, or multiplications may be performed in any order. Thus, in

$$4 + 2 + 7 + 5 = 18$$

$$\text{or } 100 - 20 - 10 - 3 = 67$$

$$\text{or } 4 \times 2 \times 7 \times 5 = 280$$

the numbers may be combined in any order desired. For example, they may be grouped easily to give

$$6 + 12 = 18$$

$$\text{and } 97 - 30 = 67$$

$$\text{and } 40 \times 7 = 280$$

A series of divisions should be taken in the order written. Thus, in $100 \div 10 \div 2$, $100 \div 10 = 10$ and then $10 \div 2 = 5$.

In a series of mixed operations, perform multiplication and divisions in order from left to right, then perform additions and subtractions in order from left to right. For example, $100 \div 4 \times 5 = 25 \times 5 = 125$, and $60 - 25 \div 5 = 60 - 5 = 55$. Now consider

$$\begin{aligned}
&60 - 25 \div 5 + 15 - 100 + 4 \times 10 \\
=\ &60 - 5 + 15 - 100 + 4 \times 10 \\
=\ &60 - 5 + 15 - 100 + 40 \\
=\ &115 - 105 \\
=\ &10
\end{aligned}$$

## Multiples and Fractions

Any number that is exactly divisible by a given number is a *multiple* of the given number. For example, 24 is a multiple of 2, 3, 4, 6, 8, and 12, since it is divisible by each of these numbers. Saying that 24 is a multiple of

3, for instance, is equivalent to saying that 3 multiplied by some whole number will give 24. Any number is a multiple of itself and also of 1.

Any number that is a multiple of 2 is an *even number*. The even numbers begin with 2 and progress by 2s as follows: 2, 4, 6, 8, 10, 12, . . . Any number that is not a multiple of 2 is an *odd number*. The odd numbers begin with 1 and progress by 2s, as follows: 1, 3, 5, 7, 9, 11, 13, . . .

Any number that can be divided into a given number without a remainder is a *factor* of the given number. The given number is a multiple of any number that is one of its factors. For example, 2, 3, 4, 6, 8, and 12 are factors of 24. The following four equalities show various combinations of the factors of 24:

$$24 = 24 \times 1 \qquad 24 = 8 \times 3$$
$$24 = 12 \times 2 \qquad 24 = 6 \times 4$$

If the number 24 is factored as completely as possible, it assumes the form
$24 = 2 \cdot 2 \cdot 2 \cdot 3$

**Zero As A Factor.**    If any number is multiplied by zero, the product is zero. For example, 5 times zero equals zero and may be written $5(0) = 0$. The zero factor law tells us that, if the product of two or more factors is zero, at least one of the factors must be zero.

**Prime Factors.**    A number that has factors other than itself and 1 is a *composite number*. For example, the number 15 is composite. It has factors 3 and 5.

A number that has no factors except itself and 1 is a *prime number*. Since it is sometimes advantageous to separate a composite number into prime factors, it is helpful to be able to recognize a few prime numbers quickly. The following series shows all the prime numbers up to 60: 2, 3, 5, 7, 11, 13, 17, 19, 23, 29, 31, 37, 41, 43, 47, 53, 59.

Notice that 2 is the only even prime number. All other even numbers are divisible by 2. Notice also that 51, for example, does not appear in the series, since it is a composite number equal to $3 \times 17$.

If a factor of a number is prime, it is called a *prime factor*. To separate a number into prime factors, begin by taking out the smallest factor. If the number is even, take out all the 2s first, then try 3 as a factor, and so forth. Thus, we have the following example:

$$540 = 2 \times 270$$
$$= 2 \times 2 \times 135$$
$$= 2 \times 2 \times 3 \times 45$$
$$= 2 \times 2 \times 3 \times 3 \times 15$$
$$= 2 \times 2 \times 3 \times 3 \times 3 \times 5$$

Since 1 is an understood factor of every number, we do not waste space recording this as one of the factors in a presentation of this kind.

A convenient way of keeping track of the prime factors is in the short division process as follows:

$$2/540$$
$$2/270$$
$$3/135$$
$$3/45$$
$$3/15$$
$$5/5$$
$$1$$

If a number is odd, its factors will be odd numbers. To separate an odd number into prime factors, take out the 3s first, if there are any. Then try 5 as a factor, and so forth. As an example:

$$5{,}775 = 3 \times 1{,}925$$
$$= 3 \times 5 \times 385$$
$$= 3 \times 5 \times 5 \times 77$$
$$= 3 \times 5 \times 5 \times 7 \times 11$$

***Tests for Divisibility.***    It is often useful to be able to tell by inspection whether a number is exactly divisible by one or more of the digits from 2 through 9. An expression which is frequently used, although it is sometimes misleading, is "evenly divisible." This expression has nothing to do with the concept of even and odd numbers, and it probably should be avoided in favor of the more descriptive expression, "exactly divisible." For the remainder of this discussion, the word *divisible* has the same meaning as "exactly divisible.'' Several tests for divisibility are listed in the following paragraphs:

1. A number is divisible by 2 if its right-hand digit is even.

2. A number is divisible by 3 if the sum of its digits is divisible by 3. For example, adding the digits of the number 6,561 produces the sum 18. Since 18 is divisible by 3, we know 6,561 is divisible by 3.

3. A number is divisible by 4 if the number formed by the two right-hand digits is divisible by 4. For example, the two right-hand digits of the number 3,524 form the number 24. Since 24 is divisible by 4, we know that 3,524 is divisible by 4.

4. A number is divisible by 5 if its right-hand digit is 5 or 0.

5. A number is divisible by 6 if it is even and the sum of its digits is divisible by 3. For example, the sum of the digits 64,236 is 21, which is divisible by 3. Since 64,236 is also an even number, we know that it is divisible by 6.

6. No short method has been found for determining whether a number is divisible by 7.

7. A number is divisible by 8 if the number formed by the three right-hand digits is divisible by 8. For example, the three right-hand digits of the number 54,272 form the number 272, which is divisible by 8. Therefore, we know that 54,272 is divisible by 8.

8. A number is divisible by 9 if the sum of its digits is divisible by 9. For example, the sum of the digits 546,372 is 27, which is divisible by 9. Therefore we know that 546,372 is divisible by 9.

# SIGNED NUMBERS

The positive numbers with which we have worked previously are not sufficient for every situation. For example, a negative number results in the operation of subtraction when the subtrahend is larger than the minuend.

## Negative Numbers

When the subtrahend happens to be larger than the minuend, this fact is indicated by placing a minus sign in front of the difference, as in the following:

$$12 - 20 = -8$$

The difference, $-8$, is said to be *negative*. A number preceded by a minus sign is a *negative number*. The number $-8$ is read "minus eight." Such a number might arise when we speak of temperature changes. If the temperature was 12 degrees yesterday and dropped 20 degrees today, the reading today would be $12 - 20$, or $-8$ degrees.

Numbers that show either a plus or a minus sign are called *signed numbers*. An unsigned number is understood to be positive and is treated as though there were a plus sign preceding it.

If it is desirable to emphasize the fact that a number is positive, a plus sign is placed in front of the number, as in $+5$, which is read "plus five." Therefore, either $+5$ or 5 indicates that the number 5 is positive. If a number is negative, a minus sign must appear in front of it, as in $-9$.

In dealing with signed numbers it should be emphasized that plus and minus signs have two separate and distinct functions. They may indicate whether a number is positive or negative, or they may indicate the operation of addition or subtraction.

When operating entirely with positive numbrs, it is not necessary to be concerned with this distinction since plus or minus signs indicate only addition or subtraction. However, when negative numbers are also involved in a computation, it is important to distinguish between a sign of operation and the sign of a number.

**Direction of Measurement.**   Signed numbers provide a convenient way of indicating opposite directions with a minimum of words. For example, an altitude of 20 ft above sea level could be designated as $+20$ ft. The same distance below sea level would then be designated as $-20$ ft. One of the most common devices utilizing signed numbers to indicate direction is the thermometer.

*Thermometer.*   The Celsius (Centigrade) thermometer shown in Figure 1-4 illustrates the use of positive and negative numbers to indicate direction of travel above and below 0. The 0 mark is the changeover point, at which the signs of the scale numbers change from $-$ to $+$.

When the thermometer is heated by the surrounding air or by a hot liquid in which it is placed, the mercury expands and travels up the tube. After the expanding mercury passes 0, the mark at which it comes to rest is read as a positive temperature. If the thermometer is allowed to cool, the mercury contracts. After passing the 0 in its downward movement, any mark at which it comes to rest is read as a negative temperature.

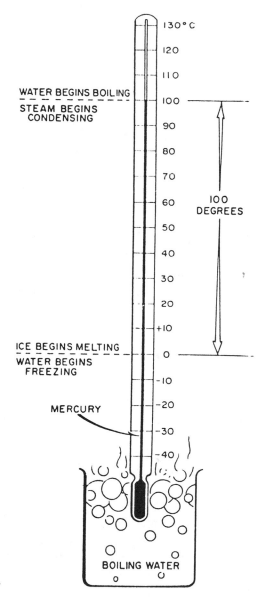

**FIGURE 1-4.**
Celsius (Centigrade) temperature scale

*Rectangular Coordinate System.* As a matter of convenience, mathematicians have agreed to follow certain conventions as to the use of signed numbers in directional measurement. For example, in Figure 1–5 a direction to the right along the horizontal line is positive, while the opposite direction (toward the left) is negative. On the vertical line, direction upward is posi-

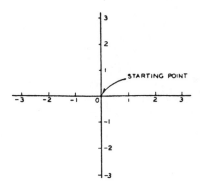

**FIGURE 1-5.**
Rectangular coordinate system

tive, while direction downward is negative. A distance of $-3$ units along the horizontal line indicates a measurement of 3 units to the left of starting point 0. A distance of $-3$ units on the vertical line indicates a measurement of 3 units below the starting point.

The two lines of the rectangular coordinate system which pass through the 0 position are the vertical axis and horizontal axis. Other vertical and horizontal lines may be included, forming a grid. When such a grid is used for the location of points and lines, the resulting "picture" containing points and lines is called a *graph*.

***The Number Line.***    Sometimes it is important to know the relative greatness (magnitude) of positive and negative numbers. To determine whether a particular number is greater than or less than another number, think of all the numbers both positive and negative as being arranged along a horizontal line. (See Figure 1–6.)

Place 0 at the middle of the line. Let the negative numbers extend from 0 toward the left. With this arrangement, positive and negative numbers are so located that they progress from smaller to larger numbers as we move from left to right along the line. Any number that lies to the left of a given number is less than the given number. This arrangement shows that any negative number is smaller than any positive number.

The symbol for *greater than* is $>$. The symbol for *less than* is $<$. It is

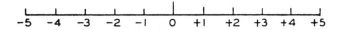

**FIGURE 1-6.**
Number line showing both positive and negative numbers

easy to distinguish between these symbols because the symbol used always opens toward the larger number. For example, "7 is greater than 4" can be written $7 > 4$ and "$-5$ is less than $-1$" can be written $-5 < -1$.

*Absolute Value.*    The *absolute value* of a number is its numerical value when the sign is dropped. The absolute value of either $+5$ or $-5$ is 5. Thus, two numbers that differ only in sign have the same absolute value.

The symbol for absolute value consists of two vertical bars placed one on each side of the number, as in $|-5| = 5$. Consider also the following:

$$|4 - 20| = 16$$
$$|+7| = |-7| = 7$$

The expression $|-7|$ is read "absolute value of minus seven."

When positive and negative numbers are used to indicate direction of measurement, we are concerned only with absolute value if we wish to know only the distance covered. For example, in Figure 1-5, if an object moves to the left from the starting point indicated by $-2$, the actual distance covered is 2 units. We are concerned only with the fact that $-2 = 2$, if our only interest is in the distance and not the direction.

## Operating with Signed Numbers

The number line can be used to demonstrate addition of signed numbers. Two cases must be considered; namely, adding numbers with like signs and adding numbers with unlike signs.

**Adding with Like Signs.**    As an example of adding with like signs, suppose that we use the number line to add $2 + 3$. (See Figure 1-7.) Since these are signed numbers, we indicate this addition as $(+2) + (+3)$. This emphasizes that among the three $+$ signs shown, two are number signs and one is a sign of operation. Line (a) above the number line shows this addition. Find 2 on the number line. To add 3 to it, go three units more in the positive direction and get 5.

To add two negative numbers on the number line, such as $-2$ and $-3$, find $-2$ on the number line and then go three units more in the negative direction to get $-5$, as in (b) above the number line.

Observation of the results of the foregoing operations on the number line leads us to the following conclusion, which may be stated as a law: *To*

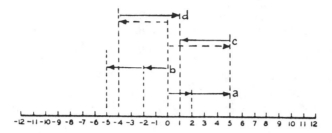

**FIGURE 1-7.**
Using the number line to add

*add numbers with like signs, add the absolute values and prefix the common sign.*

**Adding with Unlike Signs.**    To add a positive and a negative number, such as $(-4) + (+5)$, find $+5$ on the number line and go four units in the negative direction, as in line (c) above the number line in Figure 1-7. Notice that this addition could be performed in the other direction. That is, we could start at $-4$ and move 5 units in the positive direction. See line (d) in Figure 1-7.

The results of our operations with mixed signs on the number line lead to the following conclusion, which may be stated as a law: *To add numbers with unlike signs, find the difference between their absolute values and prefix the sign of the numerically greater number.*

The following examples show the addition of the numbers 3 and 5 with the four possible combinations of signs:

$$
\begin{array}{rrrr}
3 & -3 & 3 & -3 \\
5 & -5 & -5 & 5 \\
\hline
8 & -8 & -2 & 2
\end{array}
$$

In the first example, 3 and 5 have like signs and the common sign is understood to be positive. The sum of the absolute values is 8 and no sign is prefixed to this sum, thus signifying that the sign of the 8 is understood to be positive.

In the second example, the 3 and 5 again have like signs, but their common sign is negative. The sum of the absolute values is 8, and this time the common sign is prefixed to the sum. The answer is thus $-8$.

In the third example, the 3 and 5 have unlike signs. The difference between their absolute values is 2, and the sign of the larger addend is negative. Therefore, the answer is $-2$.

In the fourth example, the 3 and 5 again have unlike signs. The difference of the absolute values is still 2, but this time the sign of the larger addend is positive. Therefore, the sign prefixed to the 2 is positive (understood) and the final answer is simply 2.

These four examples could be written in a different form, emphasizing the distinction between the sign of a number and an operational sign, as follows:

$$(+3) + (+5) = +8$$
$$(-3) + (-5) = -8$$
$$(+3) + (-5) = -2$$
$$(-3) + (+5) = +2$$

**Subtraction.**    Subtraction is the inverse of addition. When subtraction is performed, we "take away" the subtrahend. This means that whatever the value of the subtrahend, its effect is to be reversed when subtraction is indicated. In addition, the sum of 5 and $-2$ is 3. In subtraction, however, to take away the effect of the $-2$, the quality $+2$ must be added. Thus the difference between $+5$ and $-2$ is $+7$.

Keeping this idea in mind, we may now proceed to examine the various combinations of subtraction involving signed numbers. Let us first consider the four possibilities where the minuend is numerically greater than the subtrahend, as in the following examples:

| 8 | 8 | $-8$ | $-8$ |
|---|---|---|---|
| 5 | $-5$ | 5 | $-5$ |
| 3 | 13 | $-13$ | $-3$ |

We may show how each of these results is obtained by use of the number line, as shown in Figure 1-8.

In the first example, we find $+8$ on the number line, then subtract 5 by making a movement that reverses its sign. Thus, we move to the left 5 units. The result (difference) is $+3$. See line (a) in Figure 1-8.

In the second example, we find $+8$ on the number line, then subtract $(-5)$ by making a movement that will reverse its sign. Thus we move to the right 5 units. The result in this case is $+13$. See line (b) in Figure 1-8.

In the third example, we find $-8$ on the number line, then subtract 5 by making a movement that reverses its sign. Thus we move to the left 5 units. The result is $-13$. See line (c) in Figure 1-8.

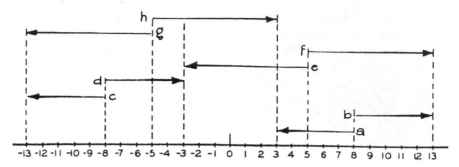

**FIGURE 1-8.**

Subtraction by use of the number line

In the fourth example, we find −8 on the number line, then reverse the sign of −5 by moving 5 units to the right. The result is −3. See line (d) in Figure 1–8.

Next, let us consider the four possibilities that arise when the subtrahend is numerically greater than the minuend, as in the following examples:

$$
\begin{array}{cccc}
5 & 5 & -5 & -5 \\
\underline{8} & \underline{-8} & \underline{8} & \underline{-8} \\
-3 & 13 & -13 & 3
\end{array}
$$

In the first example, we find +5 on the number line, then subtract 8 by making a movement that reverses its sign. Thus we move to the left 8 units. The result is −3. See line (e) in Figure 1–8.

In the second example, we find +5 on the number line, then subtract −8 by making a movement to the right that reverses its sign. The result is 13. See line (f) in Figure 1–8.

In the third example, we find −5 on the number line, then reverse the sign of 8 by a movement to the left. The result is −13. See line (g) in Figure 1–8.

In the fourth example, we find −5 on the number line, then reverse the sign of −8 by a movement to the right. The result is 3. See line (h) in Figure 1–8.

Careful study of the preceding examples leads to the following conclusion, which is stated as a law for subtraction of signed numbers: *In any subtraction problem, mentally change the sign of the subtrahend and proceed as in addition.*

**Multiplication.**   To explain the rules for multiplication of signed numbers, we recall that multiplication of whole numbers may be thought of as short-ened addition. Two types of multiplication problems must be examined: the first type involves numbers with unlike signs, and the second involves num-bers with like signs.

*Unlike Signs.*   Consider the example $3(-4)$, in which the multiplicand is negative. This means we are to add $-4$ three times; that is, $3(-4)$ is equal to $(-4) + (-4) + (-4)$, which is equal to $-12$. For example, if we have three 4-dollar debts, we owe 12 dollars in all.

When the multiplier is negative, as in $-3(7)$, we are to take away 7 three times. Thus, $-3(7)$ is equal to $-(7) - (7) - (7)$ which is equal to $-21$. For example, if 7 shells were expended in one firing, 7 the next, and 7 the next, there would be a loss of 21 shells in all. Thus, the rule is as follows: *The product of two numbers with unlike signs is negative.*

The law of signs for unlike signs is sometimes stated as follows: *Minus times plus is minus; plus times minus is minus.* Thus a problem such as $3(-4)$ can be reduced to the following steps:

1. Multiply the signs and write down the sign of the answer before work-ing with the numbers themselves.
2. Multiply the numbers as if they were unsigned numbers.

Using the suggested procedure, the sign of the answer for $3(-4)$ is found to be minus. The product of 3 and 4 is 12, and the final answer is $-12$. When there are more than two numbers to be multiplied, the signs are taken in pairs until the final sign is determined.

*Like Signs.*   When both factors are positive, as in $4(5)$, the sign of the product is positive. We are to add $+5$ four times, as follows:

$$4(5) = 5 + 5 + 5 + 5 = 20$$

When both factors are negative, as in $-4(-5)$, the sign of the product is positive. We are to take away $-5$ four times.

$$-4(-5) = -(-5) - (-5) - (-5) - (-5)$$
$$= +5 + 5 + 5 + 5$$
$$= 20$$

Remember that taking away a negative 5 is the same as adding a positive 5. For example, suppose someone owes a man 20 dollars and pays him back (or diminishes the debt) 5 dollars at a time. He takes away a debt of 20 dollars by giving him four positive 5-dollar bills, or a total of 20 positive dollars in all.

The rule developed by the foregoing example is as follows: *The product of two numbers with like signs is positive.*

Knowing that the product of two positive numbers or two negative numbers is positive, we can conclude that the product of any even number of negative numbers is positive. Similarly, the product of any odd number of negative numbers is negative.

The laws of signs may be combined as follows: Minus times plus is minus; plus times minus is minus; minus times minus is plus; plus times plus is plus. Use of this combined rule may be illustrated as follows:

$$4(-2) \times (-5) \times (6) \times (-3) = -720$$

Taking signs in pairs, the understood plus on the 4, times the minus on the 2 produces a minus. This minus times the minus on the 5 produces a plus. This plus times the understood plus on the 6 produces a plus. This plus times the minus on the 3 produces a minus, so we know that the final answer is negative. The product of the numbers, disregarding their signs, is 720; therefore, the final answer is $-720$.

**Division.**   Because division is the inverse of multiplication, we can quickly develop the rules for division of signed numbers by adapting the multiplication rules.

Division involving two numbers with unlike signs is related to multiplication with unlike signs, as follows:

$$3(-4) = -12$$
$$-12 \div 3 = -4$$

Thus, the rule for division with unlike signs is: *The quotient of two numbers with unlike signs is negative.*

Division involving two numbers with like signs is related to multiplication with like signs, as follows:

$$3(-4) = -12$$
$$-12 \div -4 = 3$$

Thus the rule for division with like signs is: *The quotient of two numbers with like signs is positive.*

The following examples show the application of the rules for dividing signed numbers:

$$12 \div 3 = 4 \qquad -12 \div -3 = -4$$
$$-12 \div -3 = 4 \qquad 12 \div -3 = -4$$

**Special Cases.**    Two special cases arise frequently in which the laws of signs may be used to advantage. The first of such usage is in simplifying subtraction; the second is in changing the signs of the numerator when division is indicated in the form of a fraction.

*Subtraction.*    The rules for subtraction may be simplified by use of the laws of signs, if each expression to be subtracted is considered as being multiplied by a negative sign. For example, $4 - (-5)$ is the same as $4 + 5$, since minus times a minus is plus. This result also establishes a rule for the removal of parentheses.

The parentheses rule, as usually stated, is: *Parentheses preceded by a minus sign may be removed if the signs of all terms within the parentheses are changed.* This is illustrated as follows:

$$12 - (3 - 2 + 4) = 12 - 3 + 2 - 4$$

The reason for the changes of sign is clear when the negative sign preceding the parentheses is considered to be a multiplier for the whole parenthetical expression.

*Division in Fractional Form.*    Division is often indicated by writing the dividend as the numerator, and the divisor as the denominator of a fraction. In algebra, every fraction is considered to have three signs. The numerator has a sign, the denominator has a sign, and the fraction itself, taken as a whole, has a sign. In many cases, one or more of these signs will be positive, and thus will not be shown. For example, in the following fraction the sign of the numerator and the sign of the denominator are both positive (understood) and the sign of the fraction itself is negative:

$$\frac{-4}{5}$$

Fractions with more than one negative sign are always reducible to a simpler form with one negative sign. For example, the sign of the numerator and the sign of the denominator may be both negative. We note that minus divided by minus gives the same result as plus divided by plus. Therefore, we may change to the less complicated form having plus signs (understood) for both numerator and denominator, as follows:

$$\frac{-15}{-5} = \frac{+15}{+5} = \frac{15}{5}$$

Since $-15$ divided by $-5$ is 3, and 15 divided by 5 is also 3, we conclude that the change of sign does not alter the final answer. The same reasoning may be applied in the following example, in which the sign of the fraction itself is negative:

$$-\left(\frac{-15}{-5}\right) = -\left(\frac{+15}{+5}\right) = -\left(\frac{15}{5}\right)$$

When the fraction itself has a negative sign, as in this example, the fraction may be enclosed in parentheses temporarily, for the purpose of working with the numerator and denominator only. Then the sign of the fraction is applied separately to the result, as follows:

$$-\left(\frac{-15}{-5}\right) = -\left(\frac{-15}{-5}\right) = -(3) = -3$$

All of this can be done mentally.

If a fraction has a negative sign in one of the three sign positions, this sign may be moved to another position. Such an adjustment is an advantage in some types of complicated expressions involving fractions. An example of this type of sign change is as follows:

$$-\left(\frac{15}{5}\right) = \frac{-15}{5} = \frac{15}{-5}$$

In the first expression of the foregoing example, the sign of the numerator is positive (understood) and the sign of the fraction is negative. Changing both of these signs, we obtain the second expression. To obtain the third expression from the second, we change the sign of the numerator and the sign of the denominator. Observe that the sign changes in each case involve a pair of signs. This leads to the law of signs for fractions: *Any two of the three signs of a fraction may be changed without altering the value of the fraction.*

## Axioms and Laws

An axiom is self-evident truth. It is a truth that is so universally accepted that it does not require proof. For example, the statement that a "straight line is the shortest distance between two points" is an axiom from plane geometry. One tends to accept the truth of an axiom without proof, because everything which is axiomatic is, by its very nature, obviously true. On the other hand, a law (in the mathematical sense) is the result of defining certain quantities and relationships, and then developing logical conclusions from the definitions.

**Axioms of Equality.** The four axioms of equality that we are concerned with in arithmetic and algebra are stated as follows:

*Axiom Number 1.* When the same quantity is added to each of two equal quantities, the resulting quantities are equal. This is sometimes stated as follows: *If equals are added to equals, the results are equal.* For example, by adding the same quantity (3) to both sides of the following equation, we obtain two sums which are equal.

$$-2 = -3 + 1$$
$$-2 + 3 = -3 + 1 + 3$$
$$1 = 1$$

*Axiom Number 2.* If the same quantity is subtracted from each of two equal quantities, the resulting quantities are equal. This is sometimes stated as follows: *If equals are subtracted from equals, the results are equal.* For example, by subtracting 2 from both sides of the following equation we obtain results which are equal.

$$5 = 2 + 3$$
$$5 - 2 = 2 + 3 - 2$$
$$3 = 3$$

*Axiom Number 3.* If two equal quantities are multiplied by the same quantity, the resulting products are equal. This is sometimes stated as follows: *If equals are multiplied by equals, the products are equal.* For example, both sides of the following equation are multiplied by $-3$ and equal results are obtained.

$$5 = 2 + 3$$

$$(-3)(5) = (-3)(2 + 3)$$

$$-15 = -15$$

*Axiom Number 4.* If two quantities are divided by the same quantity, the resulting quotients are equal. This is sometimes stated as follows: *If equals are divided by equals, the results are equal.* For example, both sides of the following equation are divided by 3, and the resulting quotients are equal.

$$12 + 3 = 15$$

$$\frac{(12 + 3)}{3} = \frac{15}{3}$$

$$4 + 1 = 5$$

These axioms are especially useful when letters are used to represent numbers. If we know that $5x = -30$, for instance, then dividing both $5x$ and $-30$ by 5 leads to the conclusion that $x = -6$.

**Laws for Combining Numbers.** Numbers are combined in accordance with the following basic laws:

1. The associative laws of addition and multiplication
2. The commutative laws of addition and multiplication
3. The distributive law

*Associative Law of Addition.* The word *associative* suggests association or grouping. This law states that the sum of three or more addends is the same regardless of the manner in which they are grouped. For example, $6 + 3 + 1$ is the same as $6 + (3 + 1)$ or $(6 + 3) + 1$.

This law can be applied to subtraction by changing signs in such a way that all negative signs are treated as number signs rather than operational signs. That is, some of the addends can be negative numbers. For example, $6 - 4 - 2$ can be written as $6 + (-4) + (-2)$. By the associative law, this is the same as

$$6 + [(-4) + (-2)] \text{ or } [6 \text{ or } (-4)] + (-2)$$

However, $6 - 4 - 2$ is not the same as $6 - (4 - 2)$; the terms must be expressed as addends before applying the associative law of addition.

*Associative Law of Multiplication.*    This law states that the product of three or more factors is the same regardless of the manner in which they are grouped. For example, $6 \times 3 \times 2$ is the same as $(6 \times 3) \times 2$ or $6 \times (3 \times 2)$. Negative signs require no special treatment in the application of this law. For example, $6 \times (-4) \times (-2)$ is the same as $[6 \times (-4)] \times (-2)$ or $6 \times [(-4) \times (-2)]$.

*Commutative Law of Addition.*    The word *commute* means to change, substitute or move from place to place. The commutative law of addition states that the sum of two or more addends is the same regardless of the order in which they are arranged. For example, $4 + 3 + 2$ is the same as $4 + 2 + 3$ or $2 + 4 + 3$.

This law can be applied to subtraction by changing signs so that all negative signs become number signs and all signs of operation are positive. For example, $5 - 3 - 2$ is changed to $5 + (-3) + (-2)$, which is the same as $5 + (-2) + (-3)$ or $(-3) + 5 + (-2)$.

*Commutative Law of Multiplication.*    This law states that the product of two or more factors is the same regardless of the order in which the factors are arranged. For example, $3 \times 4 \times 5$ is the same as $5 \times 3 \times 4$ or $4 \times 3 \times 5$. Negative signs require no special treatment in the application of this law. For example, $2 \times (-4) \times (-3)$ is the same as $(-4) \times (-3) \times 2$ or $(-3) \times 2 \times (-4)$.

*Distributive Law.*    This law combines the operations of addition and multiplication. The word *distributive* refers to the distribution of a common multiplier among the terms of an additive expression. For example,

$$2(3 + 4 + 5) = 2 \times 3 + 2 \times 4 + 2 \times 5$$
$$= 6 + 8 + 10$$

To verify the distributive law, we note that $2(3 + 4 + 5)$ is the same as $2(12)$ or 24. Also, $6 + 8 + 10$ is 24. For application of the distributive law where negative signs appear, the following procedure is recommended:

$$3(4 - 2) = 3[4 + (-2)]$$
$$= 3(4) + 3(-2)$$
$$= 12 - 6$$
$$= 6$$

## COMMON FRACTIONS

The emphasis in this chapter so far has been on integers (whole numbers). In this section we will direct our attention to numbers which are not integers. The simplest type of number other than an integer is a *common fraction.*

Fractions are as useful as whole numbers in boiler operation because almost nothing happens on a whole hour, gallon, pound, or anything.

Example

A boiler uses 21 2/3 gallons of fuel oil per hour of normal operation. How many hours will the boiler operate on 5,000 gallons of fuel oil?

$$\text{Total hours of operation} = \text{Tank capacity} \div \text{Gallons used per hour}$$

$$= 5000 \div 21.66 \text{ (Approx)}$$

$$= 230.8 \text{ hours or } 230 \text{ hours and } 48 \text{ minutes}$$

Common fractions and integers together comprise a set of numbers called the *rational numbers.* This set is a subset of the set of real numbers.

The number line may be used to show the relationship between integers and fractions. For example, if the interval between 0 and 1 is marked off to form three equal spaces (thirds), then each space so formed is one-third of the total interval. If we move along the number line from 0 toward 1, we will have covered two of the three thirds when we reach the second mark. Thus the position of the second mark represents the number 2/3. (See Figure 1–9.)

The numerals 2 and 3 in the fraction 2/3 are names so that we may distinguish between them; 2 is the *numerator* and 3 is the *denominator.* In general, the numeral above the dividing line in a fraction is the numerator and the numeral below the dividing line is the denominator. The numerator and the denominator are the *terms* of the fraction. The word *numerator* is related to the word *enumerate.* To enumerate means to "tell how many";

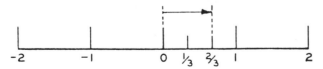

**FIGURE 1-9.**
Integers and fractions on the number line

thus the numerator tells us how many fractional parts we have in the indicated fraction. To denote means to "give a name" or "tell what kind"; thus the denominator tells us what kind of parts we have (halves, thirds, fourths, and so forth).

Attempts to define the word *fraction* in mathematics usually result in a statement similar to the following: A fraction is an indicated division. Any division may be indicated by placing the dividend over the divisor and drawing a line between them. By this definition, any number which can be written as the ratio of two integers (one integer over the other) can be considered a fraction. This leads to a further definition: Any number which can be expressed as the ratio of two integers is a rational number. Notice that every integer is a rational number, because we can write any integer as the numerator of a fraction with 1 as its denominator. For example, 5 is the same as 5/1. It should be obvious from the definition that every common fraction is also a rational number.

## Types of Fractions

Fractions are often classified as proper or improper. A proper fraction is one in which the numerator is numerically smaller than the denominator. An improper fraction has a numerator which is larger than its denominator.

**Mixed Numbers.**   When the denominator of an improper fraction is divided into its numerator, a remainder is produced along with the quotient, unless the numerator happens to be an exact multiple of the denominator. For example, 7/5 is equal to 1 plus a remainder of 2. This remainder may be shown as a dividend with 5 as its divisor, as follows:

$$\frac{7}{5} = 5 + \frac{2}{5} = 1 + \frac{2}{5}$$

The expression 1 + 2/5 is a *mixed number.* Mixed numbers are usually written without showing the plus sign; that is, 1 + 2/5 is the same as 1 2/5. When a mixed number is written as 1 2/5, care must be taken to insure that there is a space between the 1 and the 2; otherwise, 1 2/5 might be taken to mean 12/5.

**Measurement Fractions.**   Measurement fractions occur in the following problems.

If an individual paid $2 for a stateroom rug priced at $3 per yard, how many yards did he/she buy? If $6 had been spent we could find the number of yards by simply dividing the cost per yard into the amount spent. Since 6 ÷ 3 is 2, two yards could be bought for $6. The same reasoning applies when the amount spent is $2 but in this case we can only indicate the amount purchased as the indicated division 2/3. Figure 1–10 is a diagram of both the $6 purchase and the $2 purchase.

**Partitive Fractions.**    The difference between measurement fractions and partitive fractions is explained as follows: Measurement fractions result when we determine how many pieces of a given size can be cut from a larger piece. Partitive fractions result when we cut a number of pieces of equal size from a large piece and then determine the size of each smaller piece. For example, if 4 equal lengths of pipe are to be cut from a 3-foot pipe, what is the size of each piece? If the problem had read that 3 equal lengths were to be cut from a 6-foot pipe, we could find the size of each pipe by dividing the number of equal lengths into the overall length. Thus, since 6 ÷ 3 is 2, each piece would be 2 feet long. By this same reasoning, in the example we divide the overall length by the number of parts to get the size of the individual pieces; that is, 3/4 foot. The partitioned 6-foot and 3-foot pipes are shown in Figure 1–11.

**Expressing Relationships.**    When a fraction is used to express a relationship, the numerator and denominator take on individual significance. In this frame of reference, 3/4 means 3 out of 4, or 3 parts in 4, or the ratio of 3

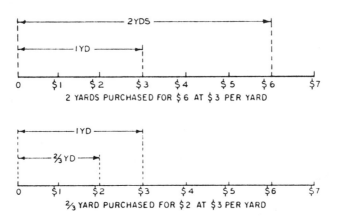

**FIGURE 1-10.**
Measurement fractions

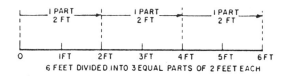

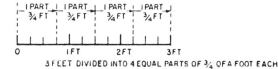

**FIGURE 1-11.**

Partitive fractions

to 4. For example, if 1 out of 3 of the men in a company are temporarily laid off, then it would be correct to state that 1/3 of the men are laid off. Observe that neither of these ways of expressing the relationship tells us the actual number of men; the relationship itself is the important thing.

## Equivalent Fractions

It will be recalled that any number divided by itself is 1. For example, 1/1, 2/2, 3/3, 4/4, and all other numbers formed this way, have the value of 1. Furthermore, any number multiplied by 1 is equivalent to the number itself. For example, 1 times 2 is 2, 1 times 3 is 3, 1 times 1/2 is 1/2, etc.

These facts are used in changing the form of a fraction to an equivalent form which is more convenient for use in a particular problem. For example, if 1 in the form 2/2 is multiplied by 3/5, the product will still have a value of 3/5 but will be in a different form, as follows:

$$\frac{2}{2} \times \frac{3}{5} = \frac{2 \times 3}{2 \times 5} = \frac{6}{10}$$

Figure 1–12 shows that 3/5 of line (a) is equal to 6/10 of line (b) where line (a) equals line (b). Line (a) is marked off in tenths. It can be readily seen that 6/10 and 3/5 measure distances of equal length.

The markings on a ruler show equivalent fractions. The major division of an inch divides it into two equal parts. One of these parts represents 1/2. The next smaller markings divide the inch into four equal parts. It will be noted that two of these parts represent the same distance as 1/2; that is, 2/4. Also, the next smaller markings break the inch into 8 equal parts. How many of these parts are equivalent to 1/2 inch? The answer is found by noting that 4/8 equals 1/2.

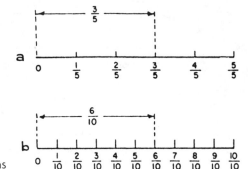

**FIGURE 1-12.**
Equivalent fractions

**Fundamental Rule of Fractions.**   The foregoing results are combined to form the fundamental rule of fractions, which is stated as follows: Multiplying or dividing both terms of a fraction by the same number does not change the value of the fraction. This is one of the most important rules in dealing with fractions.

The following examples show how the fundamental rule is used.

Example

Change 1/4 to twelfths. This problem is set up as follows:

$$\frac{1}{4} = \frac{?}{12}$$

The first step is to determine how many 4s are contained in 12. The answer is 3, so we know that the multiplier for both terms of the fraction is 3, as follows:

$$\frac{3}{3} \times \frac{1}{4} = \frac{3}{12}$$

Example

What fraction with a numerator of 6 is equal to 3/4?

*Solution:*

$$\frac{6}{?} \times \frac{3}{4}$$

We note that 6 contains 3 twice; therefore we need to double the numerator of the right-hand fraction to make it equivalent to the numerator of the fraction that we seek. We multiply both terms of 3/4 by 2, obtaining 8 as the denominator of the new fraction, as follows:

$$\frac{6}{8} = \frac{3}{4} \times \frac{2}{2}$$

Example

Change 6/16 to eighths.

*Solution:*

$$\frac{6}{16} = \frac{?}{8}$$

We note that the denominator of the fraction which we seek is 1/2 as large as the denominator of the original fraction. Therefore the new fraction may be formed by dividing both terms of the original fraction by 2, as follows:

$$\frac{6 \div 2}{16 \div 2} = \frac{3}{8}$$

**Reduction to Lowest Terms.**    It is frequently desirable to change a fraction to an equivalent fraction with the smallest possible terms; that is, with the smallest possible numerator and denominator. This process is called *reduction*. Thus, 6/30 reduced to lowest terms is 1/5. Reduction can be accomplished by finding the largest factor that is common to both the numerator and denominator and dividing both of these terms by it. Dividing both terms of the preceding example by 6 reduces the fraction to lowest terms. In computations, fractions should usually be reduced to lowest terms where possible.

If the greatest common factor cannot be found, any common factor may be removed and the process repeated until the fraction is in lowest terms: Thus, 18/48 could first be divided by 2 and then by 3.

$$\frac{18 \div 2}{48 \div 2} = \frac{9}{24} \qquad\qquad \frac{9 \div 3}{24 \div 3} = \frac{3}{8}$$

**Improper Fractions.**  Although the improper fraction is really quite "proper" mathematically, it is usually customary to change it to a mixed number. A recipe may call for 1 1/2 cups of milk, but would not call for 3/2 cups of milk.

Since a fraction is an indicated division, a method is already known for reduction of improper fractions to mixed numbers. The improper fraction 8/3 may be considered as the division of 8 by 3. This division is carried out as follows:

$$\begin{array}{r} 2 \text{ R } 2 = 2 \ 2/3 \\ 3\overline{\smash{)}8} \\ \underline{6} \\ 2 \end{array}$$

The truth of this can be verified another way: If 1 equals 3/3, then 2 equals 6/3. Thus,

$$2\frac{2}{3} = 2 + \frac{2}{3} = \frac{6}{3} + \frac{2}{3} = \frac{8}{3}$$

These examples lead to the following conclusion, which is stated as a rule: *To change an improper fraction to a mixed number, divide the numerator by the denominator and write the fractional part of the quotient in lowest terms.*

**Operating with Mixed Numbers.**  In computation, mixed numbers are often unwieldy. As it is possible to change any improper fraction to a mixed number, it is likewise possible to change any mixed number to an improper fraction. The problem can be reduced to the finding of an equivalent fraction and a simple addition.

Example

Change 2 1/5 to an improper fraction.

*Solution: Step 1*    Write 2 1/5 as a whole number plus a fraction 2 + 1/5.

*Step 2*    Change 2 to an equivalent fraction with a denominator of 5 as follows:

$$\frac{2}{1} = \frac{?}{5} \quad \frac{2(5)}{1(5)} = \frac{10}{5}$$

*Step 3*

$$\text{Add} \quad \frac{10}{5} + \frac{1}{5} = \frac{11}{5}$$

Thus,

$$2\frac{1}{5} = \frac{11}{5}$$

Example

Write 5 2/9 as an improper fraction.

*Solution:*

$$5\frac{2}{9} = 5 + \frac{2}{9}$$

$$\frac{5}{1} = \frac{?}{9}$$

$$\frac{5(9)}{1(9)} = \frac{45}{9}$$

$$\frac{45}{9} + \frac{2}{9} = \frac{47}{9}$$

Thus,

$$5\frac{2}{9} = \frac{47}{9}$$

In each of theses examples, notice that the multiplier used in Step 2 is the same number as the denominator of the fractional part of the original mixed number. This leads to the following conclusion, which is stated as a rule: *To change a mixed number to an improper fraction, multiply the whole number by the denominator of the fraction and add the numerator.* (Its denominator is the same as the denominator of the fractional part of the original mixed number.)

**Negative Fractions.**   A fraction preceded by a minus sign is negative. Any negative fraction is equivalent to a positive fraction multiplied by $-1$. For example, $-(2/5) = -1(2/5)$. The number $-(2/5)$ is read "minus two-fifths."

We know that the quotient of two numbers with unlike signs is negative. Therefore, $-(2/5) = -(2/5)$ and $2/(-5) = -(2/5)$. This indicates that a negative fraction is equivalent to a fraction with either a negative numerator or a negative denominator.

The fraction $2/(-5)$ is read "two over minus five." The fraction $(-2)/5$ is read "minus two over five."

A minus sign in a fraction can be moved about at will. It can be placed before the numerator, before the denominator, or before the fraction itself. Thus, $(-2)/5 = 2/(-5) = -(2/5)$.

Moving the minus sign from numerator to denominator, or vice versa, is equivalent to multiplying the terms of the fraction by $-1$. This is shown in the following examples:

$$\frac{-2(-1)}{5(-1)} = \frac{2}{(-5)} \text{ and } \frac{2(-1)}{-5(-1)} = \frac{-2}{5}$$

A fraction may be regarded as having three signs associated with it—the sign of the numerator, the sign of the denominator, and the sign preceding the fraction. Any two of these signs may be changed without changing the value of the fraction. Thus,

$$-\frac{3}{4} = \frac{-3}{4} = \frac{3}{-4} = -\frac{-3}{-4}$$

## Operations with Fractions

It will be recalled from the discussion of denominate numbers that numbers must be of the same denomination to be added. We can add pounds to pounds, pints to pints, but not ounces to pints. If we think of fractions loosely as denominate numbers, it will be seen that the rule of likeness applies also to fractions. We can add eighths to eighths, fourths to fourths, but not eighths to fourths. To add 1/5 inch to 2/5 inch we simply add the numerators and retain the denominator unchanged. The denominator is fifths; as with denominate numbers, we add 1 fifth to 2 fifths to get 3 fifths, or 3/5.

**Like and Unlike Fractions.**   We have shown that like fractions are added by simply adding the numerators and keeping the denominator. Thus, 3/8 + 2/8 = (3 + 2)/8 = 5/8 or 5/16 + 2/16 = 7/16.

Similary we can subtract like fractions by subtracting the numerators, as follows: 7/8 − 2/8 = (7 − 2)/8 = 5/8.

The following examples will show that like fractions may be divided by dividing the numerator of the dividend by the numerator of the divisor.

Example

$$\left(\frac{3}{8}\right) \div \left(\frac{1}{8}\right) = ?$$

*Solution:* We may state the problem as a question: "How many times does 1/8 appear in 3/8, or how many times may 1/8 be taken from 3/8?"

$$\frac{3}{8} - \frac{1}{8} = \frac{2}{8}$$

$$\frac{2}{8} - \frac{1}{8} = \frac{1}{8}$$

$$\frac{1}{8} - \frac{1}{8} = \frac{0}{8} = 0$$

We see that 1/8 can be subtracted from 3/8 three times. Therefore,

$$\left(\frac{3}{8}\right) \div \left(\frac{1}{8}\right) = 3$$

When the denominators of fractions are unequal, the fractions are said to be unlike. Addition, subtraction, or division cannot be performed directly on unlike fractions. The proper application of the fundamental rule, however, can change their form so that they become like fractions. Then all the rules for like fractions apply.

**Lowest Common Denominator.**    To change unlike fractions to like fractions, it is necessary to find a *common denominator* and it is usually advantageous to find the *lowest common denominator* (LCD). This is nothing more than the least common multiple of the denominators.

*Least Common Multiple.*    If a number is a multiple of two or more different numbers, it is called a *common multiple.* Thus, 24 is a common multiple of 6 and 2. There are many common multiples of these numbers. The numbers 36, 48, and 54, to name a few, are also common multiples of 6 and 2.

The smallest of the common multiples of a set of numbers is called the *least common multiple.* It is abbreviated LCM. The least common multiple of 6 and 2 is 6. To find the least common multiple of a set of numbers, first separate each set into prime factors.

Suppose that we wish to find the LCM of 14, 24, and 30. Separating these numbers into prime factors we have

$$14 = 2 \times 7$$
$$24 = 2^3 \times 3$$
$$30 = 2 \times 3 \times 5$$

The LCM will contain each of the various prime factors shown. Each prime factor is used the greatest number of times that it occurs in any one of the numbers. Notice that 3, 5, and 7 each occur only once in any one number. On the other hand, 2 occurs three times in one number. We get the following result:

$$LCM = 2^3 \times 3 \times 5 \times 7$$
$$= 840$$

Thus, 840 is the least common multiple of 14, 24, and 30.

*Greatest Common Divisor.* The largest number that can be divided into each of two or more given numbers without a remainder is called the *greatest common divisor* of the given numbers. It is abbreviated GCD. It is also sometimes called the *highest common factor.*

In finding the GCD of a set of numbers, separate the numbers into prime factors just as for LCM. The GCD is the product of only those factors that appear in all of the numerators. Notice in the example of the previous section that 2 is the greatest common divisor of 12, 24, and 30.

To find the GCD of 650, 900, and 700, the procedure is as follows:

$$650 = 2 \times 5^2 \times 13$$
$$900 = 2^2 \times 3^2 \times 5^2$$
$$700 = 2^2 \times 5^2 \times 7$$
$$GCD = 2 \times 5^2 = 50$$

Notice that 2 and $5^2$ are factors of each numerator. The greatest common divisor is $2 \times 25 = 50$.

Using the LCD

Consider the example: $1/2 + 1/3$. The numbers 2 and 3 are both prime; so the LCD is 6. Therefore,

$$\frac{1}{2} = \frac{3}{6} \text{ and } \frac{1}{3} = \frac{2}{6}$$

Thus, the addition of 1/2 and 1/3 is performed as follows:

$$\frac{1}{2} + \frac{1}{3} = \frac{3}{6} + \frac{2}{6} = \frac{5}{6}$$

In the example

$$\frac{1}{5} + \frac{3}{10}$$

10 is the LCD.

Therefore,

$$\frac{1}{5} + \frac{3}{10} = \frac{2}{10} + \frac{3}{10}$$

$$= \frac{5}{10} = \frac{1}{2}$$

**Addition.**   It has been shown that in adding like fractions we add the numerators. In adding unlike fractions, the fractions must first be changed so that they have common denominators. We apply these same rules in adding mixed numbers. It will be remembered that a mixed number is an indicated sum. Thus, 2 1/3 is really 2 + 1/3. Adding can be done in any order. The following examples will show the application of these rules.

Example

$$2\frac{1}{3}$$
$$3\frac{1}{3}$$
$$\overline{5\frac{2}{3}}$$

This could have been written as follows:

$$2 + \frac{1}{3}$$

$$\frac{3 + \frac{1}{3}}{5 + \frac{2}{3} = 5\frac{2}{3}}$$

Example

$$4\frac{5}{7}$$

$$\frac{6\frac{3}{7}}{10\frac{8}{7}}$$

Here we change 8/7 to a mixed number 1 1/7. Then

$$10\frac{8}{7} = 10 + 1 + \frac{1}{7}$$

$$= 11\frac{1}{7}$$

Example
Add

$$\frac{1}{4}$$

$$\frac{2\frac{2}{3}}{}$$

We first change the factors so that they are like and have the least common denominator and then proceed as before.

$$\frac{1}{4} = \frac{3}{12}$$

$$\frac{2\frac{2}{3} = 2\frac{8}{12}}{2\frac{11}{12}}$$

Example

Add

$$4\frac{5}{8} = 4\frac{5}{8}$$

$$2\frac{1}{2} = 2\frac{4}{8}$$

$$\frac{1}{4} = \frac{2}{8}$$

$$6\frac{11}{8}$$

Since 11/8 equals 1 3/8, the final answer is found as follows:

$$6\frac{11}{8} = 6 + 1 + \frac{3}{8}$$

$$= 7\frac{3}{8}$$

Example

Find the total length of the piece of metal shown in Figure 1–13.

(A)

(B)

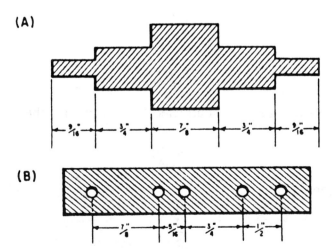

**FIGURE 1-13.**
Adding fractions to obtain total length or spacing

*Solution:*    First indicate the sum as follows:

$$\frac{9}{16} + \frac{3}{4} + \frac{7}{8} + \frac{3}{4} + \frac{9}{16} = ?$$

Changing to like fractions and adding numerators,

$$\frac{9}{16} + \frac{12}{16} + \frac{14}{16} + \frac{12}{16} + \frac{9}{16} = \frac{56}{16}$$

$$= 3\frac{8}{16}$$

$$= 3\frac{1}{2}$$

The total length is 3 1/2 inches.

**Subtraction.**   The rule of likeness applies in the subtraction of fractions as well as in addition. Some examples will show that cases likely to arise may be solved by use of ideas previously developed.

Example

Subtract 1 1/3 from 5 2/3.

$$\begin{array}{r} 5\frac{2}{3} \\ - 1\frac{1}{3} \\ \hline 4\frac{1}{3} \end{array}$$

We see that whole numbers are subtracted from whole numbers; fractions from fractions.

Example

Subtract 1/8 from 4/5.

$$\begin{array}{r} \frac{4}{5} \\ - \frac{1}{8} \\ \hline \end{array}$$

Changing to like fractions with an LCD, we have:

$$\begin{array}{r} \dfrac{32}{40} \\[6pt] -\ \dfrac{5}{40} \\ \hline \dfrac{27}{40} \end{array}$$

Example

Subtract 11/12 from 3 2/3.

$$3\,\frac{2}{3} = 3\,\frac{8}{12}$$

$$\frac{11}{12} = \frac{11}{12}$$

Regrouping 3 8/12 we have

$$2 + 1 + \frac{8}{12} = 2 + \frac{12}{12} + \frac{8}{12}$$

Then

$$3\,\frac{2}{3} = 2\,\frac{20}{12}$$

$$\frac{11}{12} = \frac{11}{12}$$

$$\overline{\phantom{xxxxxxxxxxxx}}$$

$$2\,\frac{9}{12} = 2\,\frac{3}{4}$$

Example

What is the length of the dimension marked X on the machine bolt shown in Figure 1–14(A)?

*Solution:*

$$\frac{1}{4} + \frac{1}{64} + \frac{1}{2} = \frac{16}{64} + \frac{1}{64} + \frac{32}{64} = \frac{49}{64}$$

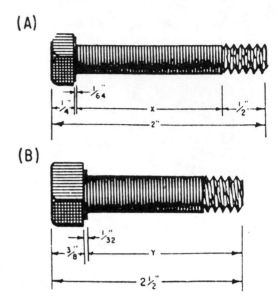

**FIGURE 1-14.**

Finding unknown dimensions by subtracting fractions

Subtract this sum from the overall length:

$$2 \quad = 1 \frac{64}{64}$$

$$\frac{49}{64} = \frac{49}{64}$$

$$1 \frac{15}{64}$$

The answer is 1 15/64 inches.

**Multiplication.**   The fact that multiplication by a fraction does not increase the value of the product may confuse those who remember the definition of multiplication presented earlier for whole numbers. It was stated that 4(5) means 5 is taken as an addend 4 times. How is it then that 1/2(4) is 2, a number less than 4? Obviously our idea of multiplication must be broadened.

Consider the following products:

$$4(4) = 16$$
$$3(4) = 12$$
$$2(4) = 8$$
$$1(4) = 4$$

$$\tfrac{1}{2}(4) = 2$$
$$\tfrac{1}{4}(4) = 1$$

Notice that as the multiplier decreases, the product decreases, until, when the multiplier is a fraction, the product is less than 4. The product continues to decrease as the fraction decreases. The fraction introduces the "part of" idea: 1/2(4) means 1/2 of 4; 1/4(4) means 1/4 of 4.

The definition of multiplication stated for whole numbers may be extended to included fractions. Since 4(5) means that 5 is to be used as an addend 4 times, we can say that with fractions the numerator of the multiplier tells how many times the numerator of the multiplicand is to be used as an addend. By the same reasoning, the denominator of the multiplier tells how many times the denominator of the multiplicand is to be used as an addend. The following examples illustrate the use of this idea.

Example

The fraction 1/12 is multiplied by the whole number 4 as follows:

$$4 \times \frac{1}{12} = \frac{4}{1} \times \frac{1}{12}$$
$$= \frac{(1 + 1 + 1 + 1)}{12}$$
$$= \frac{4}{12} = \frac{1}{3}$$

This example shows that 4 (1/12) is the same as 4(1)/12.

Another way of thinking about the multiplication of 1/12 by 4 is as follows:

$$4 \times \frac{1}{12} = \frac{1}{12} + \frac{1}{12} + \frac{1}{12} + \frac{1}{12}$$
$$= \frac{4}{12} = \frac{1}{3}$$

Example

The fraction 2/3 is multiplied by 1/2 as follows:

$$\frac{1}{2} \times \frac{2}{3} = \frac{2}{6}$$
$$= \frac{1}{3}$$

From these examples a general rule is developed: *To find the product of two or more fractions multiply their numerators together and write the result as the numerator of the product; multiply their denominators and write the result as the denominator of the product; reduce the answer to the lowest terms.*

In using this rule with whole numbers, write each whole number as a fraction with 1 as the denominator. For example, multiply 4 times 1/12 as follows:

$$4 \times \frac{1}{12} = \frac{4}{1} \times \frac{1}{12}$$
$$= \frac{4}{12} = \frac{1}{3}$$

In using this rule with mixed numbers, rewrite all mixed numbers as improper fractions before applying this rule, as follows:

$$2\frac{1}{3} \times \frac{1}{2} = \frac{7}{3} \times \frac{1}{2}$$
$$= \frac{7}{6}$$

A second method of multiplying mixed numbers makes use of the distributive law. This law states that a multiplier applied to a two-part expression is distributed over both parts. For example, to multiply 6 1/3 by 4 we may rewrite 6 1/3 as 6 + 1/3. Then the problem can be written as 4(6 + 1/3) and the multiplication proceeds as follows:

$$4\left(6 + \frac{1}{3}\right) = 24 + \frac{4}{3}$$
$$= 25 + \frac{1}{3}$$
$$= 25\frac{1}{3}$$

*Cancellation.*   Computation can be considerably reduced by dividing out or *cancelling* factors common to both the numerator and the denominator. We recognize a fraction as an indicated division. Thinking of 6/9 as an indicated division, we remember that we can simplify division by showing both dividend and divisor as the indicated products of their factors and then dividing like factors, or cancelling. Thus, $6/9 = (2/3) \times (3/3)$.

Dividing the factor 3 in the numerator by 3 in the denominator gives the following result:

$$\frac{2 \times \cancel{3}^{①}}{3 \times \cancel{3}_{①}} = \frac{2}{3}$$

This method is most advantageous when done before any other computation. Consider the example:

$$\frac{1}{3} \times \frac{3}{2} \times \frac{2}{5}$$

The product in factored form is

$$\frac{1}{3} \times \frac{3}{2} \times \frac{2}{5}$$

Rather than doing the multiplying and then reducing the result 6/30, it is simpler to cancel like factors first, as follows:

$$\frac{1 \times \cancel{3}^{1} \times \cancel{2}^{1}}{\cancel{3} \times \cancel{2} \times \cancel{5}} = \frac{1}{5}$$

Likewise,

$$\frac{\cancel{2}^{①}}{\cancel{3}_{①}} \times \frac{\cancel{6}^{①^{2}}}{\cancel{4}_{2_{①}}} \times \frac{5}{9} = \frac{5}{9}$$

Here we mentally factor 6 to the form $3 \times 2$, and 4 to the form, $2 \times 2$.

Cancellation is a valuable tool in shortening operations with fractions. The general rule may be applied to mixed numbers by simply changing them to improper fractions. Thus,

$$2\frac{1}{4} \times 3\frac{1}{3} = ?$$

$$\frac{9}{4} \times \frac{10}{3} = \frac{\cancel{9}}{\cancel{4}} \times \frac{\cancel{10}}{\cancel{3}} = \frac{15}{2}$$

**Division.**   There are two methods commonly used for performing division with fractions. One is the common denominator method and the other is the reciprocal method.

*Common Denominator Method.*   The common denominator method is an adaptation of the method of like fractions. The rule is as follows: *Change the dividend and divisor to like fractions and divide the numerator of the dividend by the numerator of the divisor.* This method can be demonstrated with whole numbers, first changing them to fractions with 1 as the denominator. For example, 12/4 can be written as follows:

$$\frac{12}{4} = \frac{\dfrac{12}{1}}{\dfrac{4}{1}}$$

$$= \frac{\dfrac{12}{1}}{\dfrac{4}{1}}$$

$$= \frac{\dfrac{12}{4}}{1}$$

$$= 3$$

If the dividend and divisor are both fractions as in 1/3 dividend by 1/4, we proceed as follows:

$$\frac{1}{3} \div \frac{1}{4} = \frac{4}{12} \div \frac{3}{12}$$

$$= \frac{4 \div 3}{12 \div 12}$$

$$= \frac{(4 \div 3)}{1}$$

$$= 4 \div 3 = 1 \frac{1}{3}$$

***Reciprocal Method.*** The word *reciprocal* denotes an interchangeable relationship. It is used in mathematics to describe a specific relationship between two numbers. We say that two numbers are reciprocals of each other if their product is one. If the example $4 \times 1/4 = 1$, the fractions 4/1 and 1/4 are reciprocals. Notice the interchangeability: 4 is the reciprocal of 1/4 and 1/4 is the reciprocal of 4.

What is the reciprocal of 3/7? It must be a number which, when multiplied by 3/7, produces the product, 1. Therefore,

$$\frac{3}{7} \times ? = 1$$

$$\frac{3}{7} \times \frac{7}{3} = 1$$

We see that 7/3 is the only number that could fulfill the requirement. Notice that the numerator and denominator of 3/7 were simply interchanged to get the reciprocal. If we know a number, we can always find its reciprocal by dividing 1 by the number. Notice this principle in the examples below.

Example

What is the reciprocal of 7?

$$1 \div 7 = \frac{1}{7}$$

*Check:*

$$\frac{7}{1} \times \frac{1}{7} = 1$$

Notice that the cancellation process in this example does not show the usual 1s which result when dividing a number into itself. For example, when 7 cancels 7, the quotient 1 could be shown beside each of the 7s. However, since 1 as a factor has the same effect whether it is written in or simply understood, the 1s need not be written.

### Example

What is the reciprocal of 3/8?

$$1 \div \frac{3}{8} = \frac{8}{8} \div \frac{3}{8}$$

$$= 8 \div 3, \text{ or } \frac{8}{3}$$

*Check:*

$$\frac{3}{8} \times \frac{8}{3} = 1$$

### Example

What is the reciprocal of 5/2?

$$1 \div \frac{5}{2} = \frac{2}{2} \div \frac{5}{2}$$

$$= 2 \div 5$$

$$= \frac{2}{5}$$

*Check:*

$$\frac{5}{2} \times \frac{2}{5} = 1$$

### Example

What is the reciprocal of 3 1/8?

$$1 \div 3 \frac{1}{8} = \frac{8}{8} \div \frac{25}{8}$$

$$= 8 \div 25$$

$$= \frac{8}{25}$$

*Check:*

$$\frac{25}{8} \times \frac{8}{25} = 1$$

The foregoing examples lead to the rule for finding the reciprocal of any number. The reciprocal of a number is the fraction formed when 1 is divided by the number. (If the final result is a whole number, it can be considered as a fraction whose denominator is 1.) A shortcut rule which is purely mechanical and does not involve reasoning may be stated as follows: *To find the reciprocal of a number, express the number as a fraction and then invert the fraction.*

When the numerator of a fraction is 1, the reciprocal is a whole number. The smaller the fraction, the greater is the reciprocal. For example, the reciprocal of 1/1,000 is 1,000.

Also the reciprocal of any whole number is a proper fraction. Thus the reciprocal of 50 is 1/50.

**Complex Fractions.** When the numerator or the denominator, or both, in a fraction are themselves composed of fractions, the resulting expression is called a complex fraction. The following expression is a complex fraction:

$$\frac{\dfrac{3}{5}}{\dfrac{3}{4}}$$

This should be read as "three-fifths over three-fourths" or "three-fifths divided by three-fourths." Any complex fraction may be simplified by writing it as a division problem, as follows:

$$\frac{3/5}{3/4} = \frac{3}{5} \div \frac{3}{4}$$
$$= \frac{3}{5} \times \frac{4}{3}$$
$$= \frac{4}{5}$$

Similarly,

$$\frac{3\,1/3}{2\,1/2} = \frac{10}{3} \div \frac{5}{2} = \frac{\overset{2}{\cancel{10}}}{3} \times \frac{2}{\cancel{5}} = \frac{4}{3} = 1\,1/3$$

Complex fractions may also contain an indicated operation in the numerator or demoninator or both. Thus,

$$\frac{\dfrac{1}{2} + \dfrac{1}{3}}{\dfrac{9}{5} + \dfrac{1}{5}}$$

is a complex fraction. To simplify such a fraction we simplify the numerator and denominator and proceed as follows:

$$\frac{1/2 + 1/3}{9/5 + 1/5} = \frac{3/6 + 2/6}{10/5} = \frac{5/6}{2}$$
$$= \frac{5}{6} \div \frac{2}{1}$$
$$= \frac{5}{6} \times \frac{1}{2}$$
$$= \frac{5}{12}$$

Mixed numbers appearing in complex fractions usually show the plus sign. Thus, $4\ 2/5 \div 7\ 1/3$ might be written,

$$\frac{4 + 2/5}{7 + 1/3}$$

Complex fractions may be used in electronics when it is necessary to find the total resistance of several resistances in parallel circuits, as shown in Figure 1-15. The rule is: *The total resistance of a parallel circuit is 1 divided by the sum of the reciprocals of the separate resistances.* Written as a formula, this produces the following expression:

$$R_t = \frac{1}{1/R_1 + 1/R_2 + 1/R_3}$$

**FIGURE 1-15.**
Application of complex fractions
in calculating electrical resistance

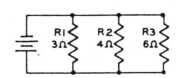

Example

Find the total resistance of the parallel circuit in Figure 1–15. Substituting the values 3, 4, and 6 for the letters $R_1$, $R_2$, and $R_3$, we have the following:

$$R_t = \frac{1}{1/3 + 1/4 + 1/6}$$

The LCD of the fractions 1/3, 1/4, and 1/6 is 12. Thus,

$$R_t = \frac{1}{4/12 + 3/12 + 2/12}$$

$$= \frac{1}{9/12}$$

(The reciprocal of $\dfrac{1}{9/12}$ is $\dfrac{12}{9/1}$ which equals 12/9.

$$= \frac{12}{9} = \frac{4}{3} = 1\frac{1}{3} \text{ ohms}$$

# DECIMALS

The origin and the meaning of the word *decimal* were discussed earlier in this chapter. The concept of place value was also discussed earlier, as well as the use of the number ten as the base for our numbering system. Another term which is frequently used to denote the base of a number system is *radix*. For example, two is the radix of the binary system and ten is the radix of the decimal system. The radix of a number system is always equal to the number of different digits used in the system. For example, the decimal system, with radix ten, has ten digits: 0 through 9.

## Decimal Fractions

A decimal fraction is a fraction whose denominator is 10 or some power of 10, such as 100, 1000, or 10,000. thus 7/10, 12/100, and 215/1000 are decimal fractions. Decimal fractions have special characteristics that make computation much simpler than with other fractions.

Decimal fractions complete the decimal system of numbers. In our study of whole numbers, we found that we could proceed to the left from the units place to tens, hundreds, thousands, and so on, indefinitely to any larger place value. Decimal fractions complete the development so that we can proceed to the right of the units place to any smaller number indefinitely.

Figure 1–16(A) shows how decimal fractions complete the system. It should be noted that as we proceed from left to right, the value of each place is one-tenth the value of the preceding place, and that the system continues uninterrupted.

Figure 1–16(B) shows the system again, this time using numbers. Notice in (A) and (B) that the units place is the center of the system and that the place values proceed to the right or left of it by powers of ten. Ten on the left is balanced by tenths on the right, hundreds by hundredths, thousands by thousandths, and so forth.

Notice that 1/10 is one place to the right of the units digit, 1/1000 is

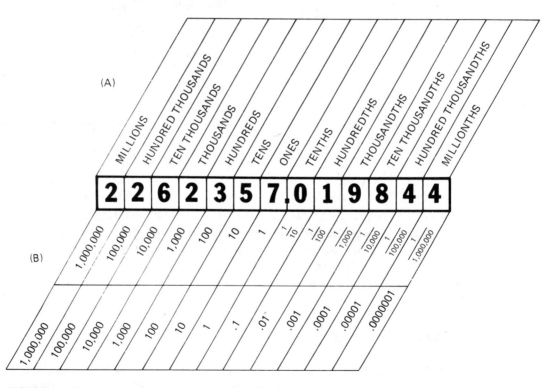

**FIGURE 1-16.**
Place values including decimals

two places to the right, etc. If a marker is placed after the units digit, we can decide whether a decimal digit is in the tenths, hundredths, or thousandths position by counting places to the right of the marker. In some European countries, the marker is a comma; but in the English-speaking countries, the marker is the *decimal point*.

Thus, 3/10 is written 0.3. To write 3/100 it is necessary to show that 3 is in the second place to the right of the decimal point, so a zero is inserted in the first place. Thus, 3/100 is written 0.03. Similarly, 3/1000 can be written by inserting zeros in the first two places to the right of the decimal point. Thus, 3/1000 is written 0.003. In the number 0.3, we say that 3 is in the first decimal place; in 0.03, 3 is in the second decimal place. Quite frequently decimal fractions are simply called decimals when written in this shortened form.

**Writing Decimals.**   Any decimal fraction may be written in the shortened form by a simple mechanical process. Simply begin at the right-hand digit of the numerator and count off to the left as many places as there are zeros in the denominator. Place the decimal point to the left of the last digit counted. The denominator may then be disregarded. If there are not enough digits, as many place-holding zeros as are necessary are added to the left of the left-hand digit in the numerator.

Thus, in 23/10000, beginning with the digit 3, we count off four places to the left, adding two 0s as we count, and place the decimal point to the extreme left. (See Figure 1–17.) Either form is read "twenty-three ten-thousandths."

When a decimal fraction is written in the shortened form, there will always be as many decimal places in the shortened form as there are zeros in the denominator of the fractional form.

Figure 1–18 shows the fraction 23458/100000 and what is meant when it is changed to the shortened form. This figure is presented to show further that each digit of a decimal fraction holds a certain position in the digit sequence and has a particular value.

By the decimal rule of fractions, it should be clear that 5/10 = 50/100 = 500/1000. Writing the same values in the shortened way, we have 0.5 =

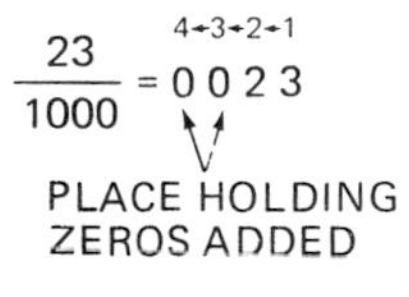

$$\frac{23}{1000} \overset{4 \cdot 3 \cdot 2 \cdot 1}{=} 0\,0\,2\,3$$

**FIGURE 1-17.**
Conversion of a decimal fraction to shortened form

PLACE HOLDING
ZEROS ADDED

$$\frac{24{,}358}{100{,}000} \quad \text{ALSO MEANS} \\ \text{THE SUM OF} \quad \begin{cases} \text{2 TENTHS} & \text{OR .2} \\ \text{4 HUNDREDTHS} & \text{OR .04} \\ \text{3 THOUSANDTHS} & \text{OR .003} \\ \text{5 TEN-THOUSANDTHS} & \text{OR .0005} \\ \text{8 HUNDRED-THOUSANDTHS} & \text{OR .00008} \\ & \overline{.24358} \end{cases}$$

**FIGURE 1-18.**
Steps in the conversion of a decimal fraction to shortened form

$0.50 = 0.500$. In other words, the value of a decimal is not changed by annexing zeros at the right-hand end of the number. This is not true of whole numbers. Thus, 0.3, 0.30, and 0.300 are equal but 3, 30, and 300 are not equal. Also notice that zeros directly after the decimal point do change values. Thus 0.3 is not equal to either 0.03 or 0.003.

Decimals such as 0.125 are frequently seen. Although the 0 on the left of the decimal point is not required, it is often helpful. This is particularly true in an expression such as $32 \div 0.1$. In this expression, the lower dot of the division symbol must not be crowded against the decimal point; the 0 serves as an effective spacer. If any doubt exists concerning the clarity of an expression such as .125, it should be written as 0.125.

**Reading Decimals.**   To read a decimal fraction in full, we read both its numerator and its denominator, as in reading common fractions. To read 0.305, we read "three hundred five thousandths." The denominator is always 1 with as many zeos as decimal places. Thus the denominator for 0.14 is 1 with two zeros, or 100. For 0.003 it is 1,000; for 0.101 it is 1,000; and for 0.3 it is 10. The denominator may also be determined by counting off place values of the decimal. For 0.13 we may think "tenths, hundredths" and the fraction is in hundredths. In the example 0.1276 we may think "tenths, hundredths, thousandths, ten-thousandths." We see that the denominator is 10,000 and we read the fraction "one thousand two hundred seventy-six ten-thousandths."

A whole number with a fraction in the form of a decimal is called a *mixed decimal*. Mixed decimals are read in the same manner as mixed numbers. We read the whole number in the usual way followed by the word *and* and then read the decimal. Thus, 160.32 is read "one hundred sixty and thirty-two hundredths." The word *and* in this case, as with mixed numbers, means plus. The number 3.2 means three plus two tenths.

It is also possible to have a complex decimal. A *complex decimal* contains a common fraction. The number 0.3 1/3 is a complex decimal and is read "three and one-third tenths." The number 0.87 1/2 means 87 1/2 hundredths. The common fraction in each case forms a part of the last or right-hand place.

In actual practice when numbers are called out for recording, the above procedure is not used. Instead, the digits are merely called out in order with the proper placing of the decimal point. For example, the number 216.003 is read, "two one six point zero zero three." The number 0.05 is read, "zero point zero five."

## Equivalent Decimals

Decimal fractions may be changed to equivalent fractions of higher or lower terms, as is the case with common fractions. If each decimal fraction is written in its common fraction form, changing to higher terms is accomplished by multiplying both numerator and denominator by 10, or 100, or some higher power of 10. For example, if we desire to change 5/10 to hundredths, we may do so by multiplying both numerator and denominator by 10. Thus, 5/10 = 50/100. In the decimal form, the same thing may be accomplished by simply annexing a zero. Thus, 0.5 = 0.50.

Annexing a 0 on a decimal has the same effect as multiplying the common fraction form of the decimal by 10/10. This is an application of the fundamental rule of fractions. Annexing two 0s has the same effect as multiplying the common fraction form by 100/100; annexing three 0s has the same effect as multiplying by 1000/1000; and so forth.

**Reduction to Lower Terms.** When dealing with decimal fractions, reducing to lower terms is known as *rounding*. If it is desired to reduce 6.3000 to lower terms, we may simply drop as many end zeros as necessary since this is equivalent to dividing both terms of the fraction by some power of ten. Thus, we see that 6.3000 is the same as 6.300, 6.30, or 6.3.

It is frequently necessary to reduce a number such as 6.427 to some lesser degree of precision. For example, suppose that 6.427 is to be rounded to the nearest hundredth. The question to be decided is whether 6.427 is closer to 6.42 or 6.43. The best way to decide this question is to compare the fractions 420/1000, 427/1000, and 430/1000. It is obvious that 427/1000 is closer to 430/1000, and 430/1000 is equivalent to 43/100; therefore we say that 6.427, correct to the nearest hundredth, is 6.43.

A mechanical rule for rounding off can be developed from the foregoing analysis. Since the digit in the tenths place is not affected when we round 6.427 to hundredths, we may limit our attention to the digits in the hundredths and thousandths places. Thus the decision is made by asking whether 27 is closer to 20 or 30. Noting that 25 is half way between 20 and 30, it is clear that anything greater than 25 is closer to 30 than it is to 20.

In any number between 20 and 30, if the digit in the thousandths place is greater than 5, then the number formed by the hundredths and thousandths digits is greater than 25. Thus we would round the 27 in our original problem to 30, as far as the hundredths and thousandths digits are concerned. This result could be summarized as follows: When rounding to hundredths, if the digit in the thousandths place is greater than 5 increase the digit in the hundredths place by 1 and drop the digit in the thousandths place.

The digit in the thousandths place may be any one of the ten digits, 0 through 9. If these ten digits are split into two groups, one composed of the five smaller digits (0 through 4) and the other composed of the five larger digits, then 5 is counted as one of the larger digits. Therefore, the general rule for rounding off is stated as follows: *If the digit in the decimal place to be eliminated is 5 or greater, increase the digit in the next decimal place to the left by 1. If the digit to be eliminated is less than 5, leave the retained digits unchanged.*

The following examples illustrate the rule for rounding off.

1. 0.1414 rounded to thousandths is 0.141.
2. 3.147 rounded to tenths is 3.1.
3. 475 rounded to the nearest hundred is 500.

Observe carefully that the answer to example 2 is not 3.2. Some trainees make the error of treating the rounding process as a kind of chain reaction, in which one first rounds 3.147 to 3.15 and then rounds 3.15 to 3.2. The error of this method is apparent when we note that 147/1000 is closer to 100/1000 than it is to 200/1000.

Problems of the following type are sometimes confusing: Reduce 2.998 to the nearest hundredth. To drop the end figure we must increase the next figure by 1. The final result is 3.00. We retain the zeros to show that the answer is carried to the nearest hundredth.

**Changing Decimals to Common Fractions.**    Any decimal may be reduced to a common fraction. To do this we simply write out the numerator and denominator in full and reduce to lowest terms. For example, to change 0.12 to a common fraction, we simply write out the fraction in full, 12/100, and reduce to lowest terms,

$$\dfrac{\overset{\textstyle\textcircled{3}}{\cancel{12}}}{\underset{\textstyle\textcircled{25}}{\cancel{100}}} = \dfrac{3}{25}$$

Likewise, 0.77 is written 77/100, but this is in lowest terms so the fraction cannot be further reduced.

One way of checking to see if a decimal fraction can be reduced to lower terms is to consider the makeup of the decimal denominator. The denominator is always 10 or a power of 10. Inspection shows that the prime factors of 10 are 5 and 2. Thus, the numerator must be divisible by 5 or 2 or both, or the fraction cannot be reduced.

Example

Change the decimal fraction 0.0625 to a common fraction and reduce to lowest terms.

*Solution:*

$$0.0625 = \frac{625}{10000}$$

$$= \frac{625}{10000} \div \frac{25}{25} = \frac{25}{400}$$

$$= \frac{1}{16}$$

Complex decimals are changed to common fractions by first writing out the numerator and denominator in full and then reducing the resulting complex fraction in the usual way. For example, to reduce 0.12 1/2, we first write

$$\frac{12\frac{1}{2}}{100}$$

Writing the numerator as an improper fraction we have (25/2)/100 and applying the reciprocal method of division, we have

$$\overset{①}{\underset{}{\frac{2\cancel{5}}{2}}} \times \frac{1}{10\cancel{0}} = \frac{1}{8}$$
$$\underset{④}{}$$

**Changing Common Fractions to Decimals.**　The only difference between a decimal fraction and a common fraction is that the decimal fraction has 1 with a certain number of zeros (in other words, a power of 10) for a denominator. Thus, a common fraction can be changed to a decimal if it can be reduced to a fraction having a power of 10 for a denominator.

If the denominator of the common fraction in its lowest terms is made up of the prime factors 2 or 5, or both, the fraction can be converted to an exact decimal. If some other prime factor is present, the fraction cannot be converted exactly. The truth of this is evident when we consider the denominator of the new fraction. It must always be 10 or a power 10, and we know the factors of such a number are always 2s and 5s.

The method of converting a common fraction to a decimal is illustrated in the following example.

Example

Convert 3/4 to a decimal.

*Solution:*

$$\frac{3}{4} = \frac{300}{400}$$
$$= \frac{300}{4} \times \frac{1}{100}$$
$$= 75 \times \frac{1}{100}$$
$$= 0.75$$

Notice that the original fraction could have been rewritten as 3000/4000, in which case the result would have been 0.750. On the other hand, if the original fraction had been rewritten as 30/40, the resulting division of 4 into 30 would not have been possible without a remainder. When the denominator of the original fraction has only 2s and 5s as factors, so that we know a remainder is not necessary, the fraction should be written with enough 0s to complete the division with no remainder.

Observation of the results of the foregoing example leads to a shortcut in the conversion method. Noting that the factor 1/100 ultimately enters the answer in the form of a decimal, we could introduce the decimal point as the final step without ever writing the fraction 1/100. Thus the procedure for changing fractions to decimals is as follows:

1. Annex enough 0s to the numerator of the original fraction so that the division will be exact (no remainder).
2. Divide the original denominator into the new numerator formed by annexing the 0s.
3. Place the decimal point in the answer so that the number of decimal places in the answer is the same as the number of 0s annexed to the original numerator.

If a mixed number in common fraction form is to be converted, convert only the fractional part and then write the two parts together. This is illustrated as follows:

$$2\frac{3}{4} = 2 + \frac{3}{4} = 2 + .75 = 2.75.$$

*Nondetermining Decimals.* As stated previously, if the denominator of a common fraction contains some prime factor other than 2 or 5, the fraction cannot be converted completely to a decimal. When such fractions are converted according to the foregoing rule, the resulting decimal would never terminate. Consider the fraction 1/3. Applying the rule, we have

$$
\begin{array}{r}
.333 \ldots \\
3\overline{)1.000} \\
9 \\
\hline
10 \\
9 \\
\hline
10 \\
9 \\
\hline
\end{array}
$$

The division will continue indefinitely. Any common fraction that cannot be converted exactly yields a decimal that will never terminate, and in which the digits recur. In the previous example, the recurring digit was 3. In the fraction 5/11, we have

$$
\begin{array}{r}
.4545 \\
11\overline{)5.0000} \\
4\,4 \\
\hline
60 \\
55 \\
\hline
50 \\
44 \\
\hline
60 \\
55 \\
\hline
\end{array}
$$

The recurring digits are 4 and 5.

When a common fraction generates such a repeating decimal, it becomes necessary to arbitrarily select a point at which to cease the repetition. This may be done in two ways. We may write a decimal fraction by rounding off at the desired point. For example, to round off the decimal generated by 1/3 to hundredths, we carry the division to thousandths, see that this figure is less than 5, and drop it. Thus, 1/3 rounded off to hundredths is 0.33. The other method is to carry the division to the desired number of decimal places and carry the remaining incomplete division as a common fraction—that is, we write the result of a complex decimal. For example, 1/3 carried to thousandths would be

$$
\frac{1}{3} = 3\overline{)1.000}
\begin{array}{r}
.333\tfrac{1}{3} \\
\end{array}
$$

$$
\begin{array}{r}
9 \\
\hline
10 \\
9 \\
\hline
10 \\
9 \\
\hline
1 \\
\end{array}
$$

## Operating with Decimals

In the study of whole numbers, it was established that units must be added to units, tens to tens, hundreds to hundreds, and so forth. For convenience in adding several numbers, units were written under units, tens under tens, and so forth. The addition of decimals is accomplished in the same manner.

**Addition.** In adding decimals, tenths are written under tenths, hundredths are written under hundredths, and so forth. When this is done, the decimal points fall in a straight line. The addition is the same as in adding whole numbers. Consider the following example:

$$
\begin{array}{r}
2.18 \\
34.35 \\
0.14 \\
\underline{4.90} \\
41.57
\end{array}
$$

Adding the first column on the right gives 17 hundredths or 1 tenth and 7 hundredths. As with whole numbers, we write the 7 under the hundredths column and add the 1 tenth in the tenths column—that is, the column of the next higher order. The sum of the tenths column is 15 tenths or 1 unit and 5 tenths. The 5 is written under the tenths column and the 1 is added in the units column.

It is evident that if the decimal points are kept in a straight line—that is, if the place values are kept in the proper columns—addition with decimals may be accomplished in the ordinary manner of addition of whole numbers. It should also be noted that the decimal point of the sum falls directly under the decimal points of the addends.

**Subtraction.** Subtraction of decimals likewise involves no new principles. Notice that the place values of the subtrahend in the following example are fixed directly under the corresponding place values of the minuend. Notice also that this causes the decimal points to be aligned and that the figures in the difference (answer) also retain the correct columnar alignment.

$$
\begin{array}{r}
45.76 \\
-\,31.87 \\
\hline
13.89
\end{array}
$$

We subtract column by column, as with whole numbers, beginning at the right.

**Multiplication.** Multiplication of a decimal by a whole number may be explained by expressing the decimal as a fraction.

Example

Multiply 6.12 by 4.

*Solution:*

$$\frac{4}{1} \times \frac{612}{100} = \frac{2448}{100} = 24.48$$

When we perform the multiplication keeping the decimal form, we have

$$\begin{array}{r} 6.12 \\ \underline{4} \\ 24.48 \end{array}$$

By common sense, it is apparent that the whole number 4 times the whole number 6, with some fraction, will yield a number in the neighborhood of 24. Hence, the placing of the decimal point is reasonable.

An explanation of several examples will reveal that the product of a decimal and a whole number has the same number of decimal places as the factor containing the decimal. Zeros, if there are any at the end of the decimal, should be rejected.

***Multiplication of Two Decimals.*** To show the rule for multiplying two decimals together, we multiply the decimal in fractional form first and then in the conventional way, as in the following example:

$$0.4 \times 0.37$$

Writing these decimals as common fractions, we have

$$\frac{4}{10} \times \frac{37}{100} = \frac{4 \times 37}{10 \times 100}$$

$$= \frac{148}{1000}$$

$$= 0.148$$

In decimal form the problem is

$$\begin{array}{r} 0.37 \\ \underline{0.4} \\ 0.148 \end{array}$$

The placing of the decimal point is reasonable, since the 4 tenths of 37 hundredths is a little less than half of 37 hundredths, or about 15 hundredths.

Consider the following example: $4.316 \times 3.4$. In the common fraction form, we have

$$\frac{4316}{1000} \times \frac{34}{10} = \frac{4316 \times 34}{1000 \times 10}$$

$$= \frac{146744}{10000}$$

$$= 14.6744$$

In the decimal form the problem is

$$\begin{array}{r} 4.316 \\ \underline{3.4} \\ 17264 \\ \underline{12948} \\ 14.6744 \end{array}$$

We note that 4 and a fraction times 3 and a fraction yields a product in the neighborhood of 12. Thus, the decimal point is in the logical place.

In the examples above it should be noted in each case that when we multiply the decimals together we are multiplying the numerators. When we place the decimal point by adding the number of decimal places in the multiplier and the multiplicand, we are in effect multiplying the denominators.

When the numbers multiplied together are thought of as the numerators, the decimal points may be temporarily disregarded and the numbers may be considered whole. This justifies the apparent disregard for place value in the multiplication of decimals. We see that the rule for multiplying decimals is only a modification of the rule for multiplying fractions.

To multiply numbers in which one or more of the factors contain a decimal, multiply as though the numbers were whole numbers. Mark off as many decimal places in the product as there are decimal places in the factors together.

*Multiplying by Powers of Ten.*    Multiplying by a power of 10 (10, 100, 1000, etc.) is done mechanically by simply moving the decimal point to the right as many places as there are zeros in the multiplier. For example, 0.00687 is multiplied by 1,000 by moving the decimal point three places to the right as follows:

$$1,000 \times 0.00687 = 6.87$$

Multiplying a number by 0.1, 0.01, 0.001, etc., is done mechanically by simply moving the decimal point to the left as many places as there are decimal places in the multiplier. For example, 348.2 is multiplied by 0.001 by moving the decimal point three places to the left as follows:

$$348.2 \times 0.001 = 0.3482$$

**Division.**   When the dividend is a whole number, we recognize the problem of division as that of converting a common fraction to a decimal. Thus in the example 5 ÷ 8, we recall that the problem could be written as

$$\frac{5000}{1000} \div 8 = \frac{500 \div 8}{1000}$$
$$= \frac{625}{1000}$$
$$= 0.625$$

This same problem may be worked by the following, more direct method:

$$\frac{5}{8} = 8\overline{)5.000} \quad \begin{array}{r} .625 \\ \end{array}$$

$$\begin{array}{r}
.625 \\
8\,\overline{)5.000} \\
4\,8 \phantom{00} \\
\hline
20 \phantom{0} \\
16 \phantom{0} \\
\hline
40 \\
40 \\
\hline
\end{array}$$

Since not all decimals generated by division terminate so early as that in the above example, if at all, how many decimals places to carry the quotient should be predetermined. If it is decided to terminate a quotient at the third decimal place, the division should be carried to the fourth place so that the correct rounding off to the third place may be determined.

When the dividend contains a decimal, the same procedure applies as when the dividend is whole. In the following examples the quotients have been rounded to three decimal places.

Example

$$6.31 \div 8$$

$$\begin{array}{r} .7887 = .789 \\ 8\overline{)6.3100} \\ \underline{5\ 6} \\ 71 \\ \underline{64} \\ 70 \\ \underline{64} \\ 60 \\ \underline{56} \\ 4 \end{array}$$

Example

$$0.0288 \div 32$$

$$\begin{array}{r} 0.0009 = 0.001 \\ 32\overline{)0.0288} \\ \underline{288} \end{array}$$

Observe that in each case (including the case where the dividend is whole), the quotient contains the same number of decimal places as the number used in the dividend. Notice also that the place values are rigid; that is, tenths in the quotient appear over tenths, hundredths over hundredths, etc.

*Decimal Divisors.*    In the foregoing examples, the divisor in each case was an integer. Division with divisors that are decimals may be accomplished by changing the divisor and dividend so that the divisor becomes a whole number.

Recalling that every division expression may be written in fraction form, we use the fundamental rule of fractions as follows: *Rewrite the division problem as a fraction. Multiply the numerator (dividend) and denominator (divisor) by 10, 100, or some higher power of 10; the power of 10 must be large enough to change the divisor to a whole number.* This rule is illustrated as follows:

$$\begin{aligned} 2.568 \div 0.24 &= \frac{2.568}{0.24} \\ &= \frac{2.568}{0.24} \times \frac{100}{100} \\ &= \frac{256.8}{24} \end{aligned}$$

Thus 2.568 divided by 0.24 is the same as 256.8 divided by 24.

From the mechancial standpoint, the foregoing rule has the effect of moving the decimal point to the right as many places as necessary to change the divisor to an integer. Therefore, the rule is sometimes stated as follows: *When the divisor is a decimal, change it to a whole number by moving the decimal point to the right. Balance the change in the divisor by moving the decimal point in the dividend an equal number of places to the right.*

The following example illustrates this version of the rule.

$$
\begin{array}{r}
9\ 1.1 \\
0.9_{\wedge}\overline{)81.9_{\wedge}9}
\end{array}
$$

The inverted v, called a caret, is used as a marker to indicate the new position of the decimal point. Notice that the decimal point in the quotient is placed immediately above the caret in the dividend. Alinement of the first quotient digit immediately above the 1 in the dividend, and the second quotient digit above the 9, assures that these digits are placed properly with respect to the decimal point.

***Dividing by Powers of 10.*** Division of any number by 10, 100, 1000, etc., is really just an exercise in placing the decimal point of a decimal fraction. Thus, $5{,}031 \div 100$ may be thought of as the decimal fraction $5031/100$; to remove the denominator, we simply count off two places from the right. Thus,

$$
\frac{5031}{100} = 50.31
$$

The following three examples illustrate this procedure:

$$
\begin{array}{rcl}
401 \div 10 &=& 40.1 \\
2 \div 1{,}000 &=& 0.002 \\
11{,}431 \div 100 &=& 114.31
\end{array}
$$

If the dividend already contains a decimal part, begin counting with the first number to the left of the decimal point. Thus, $243.6 \div 100 = 2.436$. When the decimal point is not shown in a number, it is always considered to be to the right of the right-hand digit.

Dividing by 0.1, 0.01, 0.001, etc., may also be accomplished by a simple mechanical rule. We simply begin at the position of the decimal point in the dividend and count off as many places to the right as there are decimal places in the divisor. The decimal point is then placed to the right of the last digit counted. If there are not enough digits, zeros may be added.

The foregoing rule is based on the fact that 0.1 is really 1/10, 0.01 is 1/100, 0.001 is 1/1000, etc. For example:

$$23 \div 0.1 = 23 \div \frac{1}{10}$$
$$= 23 \times 10/1$$
$$= 230$$

Notice that dividing by 0.1 is the same as multiplying by 10. Likewise,

$$234.1 \div 0.001 = 234.1 \div \frac{1}{1000}$$
$$= 234.1 \times \frac{1000}{1}$$
$$= 234,100$$

and,

$$24 \div 0.01 = 24 \div \frac{1}{100} = 24 \times \frac{100}{1} = 2,400$$

## REVIEW QUESTIONS

CHAPTER 1

1. Why do we study mathematics?
2. What are numerals?
3. What are natural numbers?
4. What is a form of coding in which each digit value depends upon its relative position to the other digits of the number?
5. In what system does each digit position in a number have ten times the value of the one on its right-hand side?
6. How is the binary system constructed?
7. In what field is the use of the binary system most common?
8. In mathematics, what does the word *set* imply?

9. What is the most fundamental set of numbers?

10. Write the set of natural numbers symbolically.

11. What is meant when ellipsis points are used in a number?

12. In actuality, what is a line in relation to points?

13. Define the process of carrying used in addition.

14. When is the process of borrowing used in subtraction?

15. What are numbers that have a unit of measure associated with them known as?

16. What must be done when calculating denominate numbers?

17. What mathematical process uses factors?

18. In multiplication, what is the answer to $3 \times 0$?

19. Divide 4 into 60 and write the answer.

20. Multiply $20 \times 100$ and write the answer.

21. In division, what is the dividend that cannot be divided evenly by the divisor known as?

22. In mathematics, when must the decimal point be used?

23. Is it possible to multiply denominate numbers by integers?

24. When multiplying a denominate number by another denominate number of the same kind, in what units will the answer be?

25. In mathematics, what is a multiple of a given number?

26. What is a number that has no factors except itself and 1 known as?

27. What are signed numbers?

28. What is the absolute value of $-6$ and $+6$?

29. If numbers with unlike signs are multiplied will the answer be positive or negative?

30. What is an axiom?

31. If we add equals to equals what will be the result?

32. What set of numbers do common fractions and integers comprise?

33. Name the two types of fractions.

34. State the fundamental rule of fractions.

35. What are the three signs that are associated with a fraction?

36. What is a common multiple?

37. When a whole number is multiplied by a fraction, is the answer larger or smaller than the original number?

38. When solving a problem, at what point is cancelling done?

39. Name the two methods used for dividing fractions.

40. What is a decimal fraction?

41. Write the decimal equivalent of 6/10.

42. When adding decimal fractions, how are the decimals positioned?

43. What is the simplest way of multiplying a number by 10?

44. What method is used in dividing a number by 10?

45. Divide 3706 by 10.

46. A boiler typically operates 16 hours per day. The boiler consumes 25 gallons of fuel oil per hour of operation. The storage tank has a capacity of 5,000 gallons. How many days will the boiler operate before the storage tank is empty?

47. An employee earns $9.65 per hour and receives time and one-half for all hours over 40 worked per week. The employee works 10 hours Monday, 11 1/2 hours Tuesday, 8 hours Wednesday, 9 1/2 hours Thursday, and 14 hours Friday. What will his total earnings be for the week?

48. During normal operation a boiler produces 5,000 pounds of steam per hour. The boiler uses 23 1/2 gallons of fuel oil for each hour of operation. The fuel oil storage tank has a 5,000 gallon capacity. How many pounds of steam can the boiler generate using a full tank of fuel oil?

49. A boiler operating on natural gas requires 14 cubic feet of air for each cubic foot of natural gas burned. The natural gas has a Btu content of 1,000 Btu per cubic foot. The boiler has a Btu input of 1,000,000 Btu per hour. How many cubic feet of air are required for each hour of boiler operation?

50. A 30-day supply of water treatment chemicals for a boiler are to be ordered. The boiler uses 1/4 gallon of chemical A for each hour of operation. The chemical is available only in 55-gallon drums. The boiler typically operates 16 hours per day and is in operation 7 days per week. How many drums of chemical A are needed?

# CHAPTER 2

# BOILER ROOM SAFETY

---

## Learning Objectives

The objectives of this chapter are to:

- Instill in the reader the necessity of boiler room safety
- Provide specific, necessary boiler room safety measures
- Provide basic steps to follow when an accident occurs
- Make boiler room operators aware of the need for safety training for personnel.

---

The old adage that "safety doesn't cost, it pays" is as true today as it was when it was first coined. Maximum safety is achieved by training, practice, and awareness. A serious accident could be disastrous to the operating engineer, to those who are working around him, to the equipment, and to the property.

It is true for the most part that accidents are caused—they don't just happen. They may be caused by negligence, ignorance, or by an action. If a bolt breaks while it is being turned, it is not because of negligence or ignorance; it broke because someone was turning the nut. Many times we must remove rusted bolts and nuts which may break during the process, but if extra care is exercised an injury can be prevented.

The safety rules which are listed in this chapter cover boiler room operation; however, the list is not complete because every boiler room will have some type of hazard that is not present in other boiler rooms. Therefore, a thorough safety training program should be given for the boiler room in which you will be working.

# ACCIDENT REPORTING

As a boiler room worker you should learn the proper procedure for reporting accidents. In many companies a severe reprimand is given for failure to report an accident, regardless of how minor it may be. Insurance companies usually make this a requirement and will not cover complications resulting from an unreported accident. They may even force the termination of an employee who fails to report an accident. Therefore, every cut or scrape, regardless of how small, should be given first aid. Splinters from wood or metal should be removed by someone from the medical field.

# GENERAL SAFETY CONSIDERATIONS

The following safety list is used by permission of the American Boiler Manufacturers Association, Arlington, Virginia, and emphasizes general safety categories.

1. Training. Employees must be trained in safety when beginning employment. This training in safety should be a continuous process for

the purpose of educating employees to recognize and to keep safety in their minds throughout their working careers. A training program should be established and maintained.

2. Manufacturer's Instructions. Equipment manufacturer's instructions should be followed.

3. Lockout and Tagout Procedures. Every plant should have a formalized lockout and tagout procedure that is strictly enforced.

4. Housekeeping. Good housekeeping is essential for safety and good plant operation. Poor housekeeping results in increased safety hazards. A clean and orderly environment will foster safety.

5. Clothing and Protective Equipment. Proper clothing should be worn at all times. Avoid loose clothing and jewelry. Protective equipment should be worn when necessary (i.e.: hard hats, respirators, ear plugs, goggles, gloves, safety shoes, etc.).

6. Remote Starting of Equipment. Much of the equipment in a power plant is started remotely and/or automatically without warning; therefore, employees must be warned to avoid equipment that can be started remotely. If work is to be done on this equipment, lockout and tagout procedures must be followed. Attach signs to the equipment such as: "Equipment can be started remotely and automatically."

7. Hot Surfaces. Many hot surfaces exist in a boiler room and even nonheated surfaces can become uncomfortably warm; therefore, employees, especially new employees, must be made aware of these conditions.

8. Unexpected Noises. A sudden and/or unexpected noise may cause employees to move involuntarily. Such reaction may result in injury. Precautions against this are hard to take but experience probably is the best teacher to prevent such inadvertent responses.

9. Unconventional Fuels. Sometimes unconventional fuels need to be burned in boilers. When this is done, particular attention should be paid to the hazards that can result. These may result from characteristics in the fuels, toxic chemicals in the fuel, and toxic chemicals produced through combustion. Persons knowledgeable in the use of such unconventional fuels should be consulted concerning the problems that may be encountered. Because of the wide variety and limited use, such fuels are not addressed in most training manuals.

The remainder of this chapter is divided into sections dealing with maintenance, fire, electrical equipment, steam leaks, fuel gas leaks, hot fly ash, oil firing, gas firing, pulverized coal firing, steam explosion, furnace explosion, and implosions. Each of these areas will be further divided into operating hazard, cause, effect, and prevention.

## Maintenance

### Operating Hazard
Equipment being serviced or repaired

Unexpected starting of remotely controlled equipment

Movement of equipment

Release of electrical energy

Release of fluid pressure

### Cause
Equipment not locked out, not tagged out, not placed in zero mechanical state or not placed in zero energy state

### Effect
Physical injury or death

### Prevention
Place the equipment in zero energy state or zero mechanical state

Establish and comply with lockout and tagout procedures

Use trained and alert personnel

Display warning signs

Use blocking devices or ties to prevent movement of the equipment

### Operating Hazard
Activities related to cleaning

### Cause
Failure to observe safety procedures applicable to maintenance cleaning

### Effect
Potential injury or death to personnel

### Prevention

Observe operating and maintenance instructions for maintenance cleaning

Observe all safety regulations and normal safety precautions

Provide a safe means of access for maintenance cleaning

Provide personnel with protective clothing and equipment

Establish a procedure to clean and remove residue (ash, soot, slag) frequently to prevent excessive accumulation

Report all unsafe conditions and/or unsafe practices

### Operating Hazard

Entering a confined or enclosed space (this includes but is not limited to furnace, drums, shell or firetube, gas passes, ducts, flues, bunkers, hoppers, and tanks)

### Cause

Extremely hazardous environment, i.e., toxic, flammable, or oxygen-deficient atmosphere, hot material, or darkness

### Effect

Potential danger to life and health

### Prevention

Observe all safety regulations and normal safety precautions

Provide sufficient ventilation to assure fresh air quality and quantity to maintain the health and safety of the personnel

Test for oxygen deficiency with field-type oxygen analyzers or other suitable devices

Provide a safe means of access to all work places

Develop stand-by emergency plans and procedures

Report all unsafe conditions and/or unsafe practices

Work with a partner

Test for toxic or flammable gases

Provide lights before entering a confined space

### Operating Hazard
Oxygen deficiency

### Cause
Confined or closed spaces

### Effect
Potential danger to life or health

### Prevention
Provide sufficient ventilation to assure fresh air quality and quantity to maintain the health and safety of the personnel

Wear approved respiratory protective equipment

Test for oxygen deficiency with field-type oxygen analyzers or other suitable devices

Observe all safety regulations and normal safety precuations

Report all unsafe conditions and/or unsafe practices

### Operating Hazard
Airborn contaminants, i.e., gases, vapors, fumes, dust, and mist

### Cause
Leakage

Inadequate ventilation or exhaust

### Effect
Potential danger to life and health

### Prevention
Operate ventilation and exhaust systems

Wear approved respiratory protective equipment

Wear protective clothing

Observe all safety regulations and normal safety precautions

Report all unsafe conditions and/or unsafe practices

### Operating Hazard
Unexpected starting of remotely controlled equipment

### Cause
Unprotected equipment
Operating and maintenance personnel not trained
No warning signs

### Effect
Physical injury or death

### Prevention
Install mechanical guards and/or equipment protection
Use warning signs
Use trained and alert personnel

### Operating Hazard
Exposed moving equipment

### Cause
Guards not installed

### Effect
Bodily injury
Dismemberment

### Prevention
Reinstall guards
Avoid loose clothes
Confine long hair

### Operating Hazard
Exposed fan blades

### Cause
Guards not in place

### Effect
Bodily injury
Dismemberment

### Prevention
Reinstall guards


### Operating Hazard
Exposed moving parts of sootblowers

### Cause
Guards not in place

### Effect
Injured or lost fingers
Bodily harm

### Prevention
Reinstall guards
Avoid loose clothing
Confine long hair


### Operating Hazard
Obstructed access

### Cause
Poor housekeeping

### Effect
Potential injury to personnel

### Prevention

Provide unobstructed access to all equipment and work places
Maintain all access ways in a clean safe condition

### Operating Hazard

Lack of access to equipment

### Cause

Access not provided

### Effect

Potential injury or death to personnel

### Prevention

Provide a safe means of access to all equipment and work places

### Operating Hazard

Accidental opening of access door

### Cause

Failure to bolt or lock door

### Effect

Potential injury to personnel

### Prevention

Bolt or lock all access doors

## Fire

### Operating Hazard

Fire

### Cause

Explosion

Electrical or mechanical failure

Improper operation of equipment

Poor housekeeping

### Effect

Potential injury or death to personnel

Potential equipment or property damage

### Prevention

Operate equipment in accordance with the manufacturer's recommended operating procedures

Conduct routine equipment maintenance

Practice good housekeeping

Report all unsafe conditions and/or unsafe practices

Train and drill operators in emergency fire fighting control and extinguishing procedures

Use fire protection systems

### Operating Hazard

Coal supply fire

### Cause

Spontaneous combustion

### Effect

Potential injury or death to personnel

Potential equipment and/or property damage

### Prevention

Operate equipment in accordance with the manufacturer's recommended operating procedures

Practice good housekeeping

Report all unsafe conditions and/or unsafe practices

Train and drill operators in emergency fire fighting control and extinguishing procedures

Use fire protection and/or inerting systems

Proper coal supply management

### Operating Hazard

Fire in pulverized fuel system

### Cause

Unground or pulverized coal in the mill system

### Effect

Potential injury or death to personnel

Potential equipment or property damage

### Prevention

Operate pulverized fuel system in accordance with the manufacturer's recommended operating procedures

Use advanced warning monitoring systems

Conduct routine maintenance cleaning

Report all unsafe conditions and/or unsafe practices

Train and drill operators in emergency fire fighting control and extinguishing procedures

Use inerting systems

Utilize the provisions and procedures outlined in the National Fire Protection Association, Inc. (NFPA) Article 85F—Pulverized Fuel Systems

### Operating Hazard

Fire in burner deck, fuel stations, other areas adjacent to the boiler

### Cause

Electrical or mechanical component failure

Fuel leaks

### Effect

Potential injury or death to personnel

Potential equipment and/or property damage

### Prevention

Operate the equipment in accordance with the manufacturer's recommended procedures

Locate fuel, combustible materials, and controls away from boiler surfaces

Report all unsafe conditions and/or unsafe practices

Train and drill operators in emergency fire fighting control and extinguishing procedures

Use fire protection systems

Repair leaks promptly

### Operating Hazard

Fire in air heater and pollution control equipment

### Cause

Unburned combustible material

### Effect

Potential injury or death to personnel

Potential equipment and/or property damage

### Prevention

Operate firing equipment to assure good combustion

Operate cleaning equipment to maintain clean surfaces

Conduct routine maintenance cleaning

Report all unsafe conditions and/or unsafe practices

Train and drill operators in emergency fire fighting control and extinguishing procedures

Use fire protection systems

## Electrical Equipment

### Operating Hazard
Exposed energized electrical wiring

### Cause
Damaged insulation or protective covering

### Effect
Electrical shock resulting in death, injury, or burns to personnel

### Prevention
Use care to prevent damaging insulation
Repair damaged insulation

### Operating Hazards
Open electrical boxes

### Cause
Failure to close box covers

### Effect
Electrical shock resulting in death, injury, or burns to personnel

### Prevention
Close boxes
Instruct personnel to keep boxes closed

### Operating Hazard
Opening switch box without turning off power

### Cause
Damaged safety catch permitting opening of box without shutting off switch
Deliberate tampering with the interlock

### *Effect*
Electrical shock resulting in death, injury, or burns to personnel

### *Prevention*
Repair safety catch
Turn off switch before opening box
Do not tamper with the interlocks

### *Operating Hazard*
Working on energized electrical equipment

### *Cause*
Second party closing switch which energizes equipment

### *Effect*
Electrical shock resulting in death, injury, or burns to personnel

### *Prevention*
Follow lockout and tagout procedures

### *Operating Hazard*
Improper use of tools and lights

### *Cause*
Lack of grounding
Ground prong cut off
Using two wire extension cords
Not grounding "cheater" plug (adapter plug)
Body contact with wet surfaces
Damaged insulation
Using lights without guards

### *Effect*
Electrical shock resulting in death, injury, or burns

### Prevention

Do not cut off ground prong

Use only properly grounded 3-wire heavy duty extension cords

Ground "cheater" plug if used

Use double insulated portable tools

Use low-voltage trouble lights or battery-operated lights

Make sure guard is installed on light

### Operating Hazard

Combustible dust entering electrical equipment

### Cause

Not keeping dust proof equipment closed

Poor housekeeping

### Effect

Death, injury, or burns to personnel

Equipment damage

Explosion and/or fire

### Prevention

Keep dust proof equipment closed

Practice good housekeeping procedures

Insure proper operation of purge equipment

## Steam Leaks

### Operating Hazard

Steam leaks

### Cause

Damaged or corroded pipes and/or other pressure parts

Invisibility of superheated steam leaks

Defective or improperly installed gasket

*Effect*

Severe burns

*Prevention*

Keep all joints and pipes tight

Warn personnel of hazards of steam leaks

Warn personnel of invisibility of superheated steam leaks

## Fuel Gas Leaks

*Operating Hazards*

Fuel gas leaks

*Cause*

Improperly assembled joints

Damage to fuel carrying piping, valves, and fittings

Leaking gasket

*Effect*

Injury and/or death to personnel resulting from explosions

Asphyxiation

Burns resulting from fires

*Prevention*

Keep all piping, valves, and fittings in good repair

Test for leaks before placing equipment in operation

Avoid using pipes for walkways or support of other equipment

Warn personnel of hazards so they will report leaks promptly

## Hot Fly Ash

*Operating Hazard*

Hot fly ash accumulations in boiler flues and plenums

### Cause

Fly ash accumulating in flues and plenums

Personnel stepping in fly ash that is still hot

Fly ash may retain heat for a number of weeks

No visible difference between hot fly ash and cold fly ash

"Quicksand" action of fly ash when stepped on

Explosive effect of water on hot fly ash

### Effect

Severe burns to legs and other parts of the body coming in contact with
hot fly ash

Overloading support system causing failure

### Prevention

Warnings to all personnel concerning this matter

Allow sufficient cooling time before walking on fly ash

Remove the hot fly ash with caution and suitable equipment

Do not spray water on hot fly ash

Probe temperature of fly ash before walking on it

# Oil Firing

### Operating Hazard

Low fuel oil temperature (on equipment burning fuel oil which requires
heating prior to combustion)

### Cause

Faulty and/or fouled heater equipment

Oil temperature control setting too low

Heater electric power off

Steam supply closed

Explosion

### Effect

Poor atomization

Dirty or smokey fire
Discharge of unburned oil into furnace
Fireside explosion or puff
Fire
Boiler damage
Property damage
Loss of life and/or injury to personnel

### Prevention
Check oil temperture periodically
Check heaters periodically
Follow the manufacturer's instructions

### Operating Hazard
High fuel oil temperature

### Cause
Improper thermostat setting
Steam control valve stuck open (steam heater)
Electric supply contacts welded closed (electric heater)
Explosion

### Effect
Poor atomization
Oil gasification
Unstable flame
Fireside explosion or puff
Fire
Boiler damage
Property damage
Loss of life and/or injury to personnel

### Prevention
Check heaters and controls periodically
Follow the manufacturer's instructions

### Operating Hazard

Low atomizing air or steam pressure

### Cause

Supply line valves inoperative or not fully open
Improper control valve setting
Low supply pressure
Explosion

### Effect

Poor atomization
Dirty or smokey fires
Discharge of unburned oil into furnace
Fireside explosion or puff
Fire
Boiler damage
Property damage
Loss of life and/or injury to personnel

### Prevention

Check supply and controls periodically
Follow the manufacturer's instructions

### Operating Hazard

Wet steam (steam atomizing)

### Cause

Steam wet from source
Steam line not insulated
Steam traps not working
Explosion

### Effect

Poor atomization
Dirty or smokey fires

Discharge of unburned oil into furnace
Fireside explosion or puff
Fire
Boiler damage
Property damage
Loss of life and/or injury to personnel

### Prevention

Insulate all steam lines
Check proper trap operation periodically
Follow the manufacturer's instructions

### Operating Hazard

Worn or damaged atomizer (sprayer plate)

### Cause

Abrasive material in oil
Normal wear
Leaving out of service burner tip in hot furnace
Tip abuse
Explosion

### Effect

Fires
Incomplete or smokey combustion
Flare back

### Prevention

Check tips regularly
Use copper tools to clean tips
Follow the manufacturer's instructions
Replace gaskets when cleaning or replacing tips
Do not use copper tools for cleaning stainless steel parts

## Gas Firing

### *Operating Hazard*
Gas line leaks

### *Cause*
Gas line damage
Excessive pressure

### *Effect*
Explosion
Fire
Asphyxiation

### *Prevention*
Protect gas piping from damage
Color code piping
Provide adequate ventilation
Provide for leak detection
Insure safety devices are operative

### *Operating Hazard*
Gas relief valve or atmospheric vent discharge

### *Cause*
Excessive pressure
Diaphragm rupture in regulators
Normal vent discharge

### *Effect*
Explosion
Fire
Loss of life and/or injury to personnel
Property damage

### Prevention
Pipe all relief valves and vents to a point of safe discharge

### Operating Hazard
Gas line repair

### Cause
Damaged piping or valves

### Effect
Explosion
Fire
Loss of life and/or injury to personnel
Property damage

### Prevention
Use accepted methods for purging and recharging gas lines
Follow the NFPA Article 54

### Operating Hazard
Wet gas

### Cause
Presence of distillate in gas

### Effect
Flame out and re-ignition
Explosion
Fire
Loss of life and/or injury to personnel
Boiler or property damage

### Prevention
Follow the NFPA Article 54 for wet gas supply system

### Operating Hazard

Significant change in Btu rating of gas

### Cause

Multiple gas sources with different heating values

### Effect

Poor combustion

Explosion

Fire

Boiler and/or property damage

Loss of life and/or injury to personnel

### Prevention

Appropriate alarms

Use combustion controls that compensate for Btu changes

### Operating Hazard

High gas pressure

### Cause

Defective gas pressure regulator

Material under seat of gas pressure regulator

Defective high-pressure switch

### Effect

Fuel rich mixture

Fireside explosion

Fire

Loss of life and/or injury to personnel

Boiler and/or property damage

### Prevention

Monitor regulator for proper operation

Check operation and setting of pressure switch periodically

Repair or replace defective regulators and switches

## Pulverized Coal Firing

### Operating Hazard

Unstable combustion resulting from low $NO_x$-control method

### Cause

Delayed air-fuel mixing to achieve flame temperatures and $NO_x$-levels can result in unstable flames

### Effect

Furnace explosion

Property damage

Physical injury or death

Forced boiler shutdown

### Prevention

Follow the manufacturer's instructions

Maintain equipment

Follow the provisions of the NFPA Article 85 E and F

### Operating Hazard

Pulverized coal leaks

### Cause

Leaking fuel piping

Broken fuel pipe

Eroded fuel pipe

Leaked coal dust not cleaned up

### Effect

Explosion

Fire

Physical injury and/or death to personnel

Forced boiler shutdown

Leaked coal dust not removed

### Prevention

Good maintenance

Good house cleaning

Follow the provision of the NFPA Article 85 E and F

### Operating Hazard

Inadequate ignition when initially lighting a burner

### Cause

Ignitor malfunctioning

### Effect

Explosion

Physical injury or death

### Prevention

Maintain ignitors

Provide operator training

Follow the provisions of NFPA Article 85 E and F

### Operating Hazard

Flame out due to inadequate ignitor energy during low load operation

### Cause

Excessive turn down

### Effect

Explosion

Physical injury and/or death to personnel

### Prevention
Provide operator training
Follow the provisions of NFPA Article 85 E and F


### Operating Hazard
Mill explosion or puff
Mill fires

### Cause
Coal hang up
Improper mill operation
Control malfunction

### Effect
Equipment damage
Physical injury and/or death to personnel

### Prevention
Eliminate foreign materials in coal feed to mill (rags, metal, rods)
Provide operator training
Provide control maintenance
Follow the provisions of NFPA Article 85 E and F


### Operating Hazard
Windbox fires
Burner fires

### Cause
Coking of burner nozzles
Improperly positioned secondary air damper vanes
Insufficient airflow in coal piping
Fuel leaks

### *Effect*

Fire spreading to building

Physical injury and/or death to personnel

### *Prevention*

Operator training

Maintenance

Routine inspections

## Steam Explosion

### *Operating Hazard*

Defective safety valves

### *Cause*

Obstruction between boiler and valve

Valve damaged or coroded (internal)

Lever tied down

Obstruction on valve outlet

### *Effect*

Will not lift to release excess pressure

Impose excess pressure on the boiler

Rupture the boiler

Loss of life and/or injury to personnel

Property damage

### *Prevention*

Replace or repair safety valve

Remove obstructions

Periodically test valve as per the American Society of Mechanical
Engineers (ASME)

### *Operating Hazard*

Defective steam pressure gauges

### Cause

Broken gauge

Gauge not in calibration

Blockage in line from boiler to gauge

Gauge cock closed

Multiple gauges not in agreement

### Effect

Gauges not showing the correct pressure

Boiler may be under excessive pressure

Prevents operator from being aware of true operating conditions

### Prevention

Calibrate gauge regularly

Replace defective gauges

Inspect gauge connection and piping to boiler for blockage and/or closed cock

### Operating Hazard

Low water

### Cause

Defective low water cutoff

Low water cutoff bypassed

Improper water column blowdown procedure

Equalizing lines restricted or plugged

Tampering with low water control

Defective boiler water feed system

Operator error

Defective or inoperative gauge glass

### Effect

Overheated boiler surfaces

Ruptured boiler

Loss of life and/or injury to personnel
Property damage

### Prevention
Verify operation of boiler water feed system periodically
Prove low water cutoff operation periodically
Use proper water column blowdown procedures
Train boiler operators
Do not tamper with low water controls
Replace defective low water control
Inspect equalizing line (especially the lower line)

### Operating Hazard
Bypassed controls

### Cause
Defective electrical wiring
Tampering with controls and electrical wiring

### Effect
Control will not function
Boiler may rupture
May cause furnace explosion
Loss of life
Property damage

### Prevention
Verify proper operation of controls periodically
Correct electrical wiring defects immediately
Do not tamper with controls

### Operating Hazard
Tampering with controls

### Cause

Deliberate action by personnel

Lack of knowledge on the part of the personnel

Inadequate training

### Effect

Improper operation of the boiler

Boiler may rupture

May cause furnace explosion

Loss of life

Property damage

### Prevention

Read and follow manufacturer's instructions

Prevent access by unauthorized personnel by locking equipment cabinet

Properly train operators

### Operating Hazard

Poor maintenance

### Cause

No definite maintenance policy and procedure

Lack of interest by the boiler owner

Poorly or inadequately trained personnel

No one assigned the maintenance responsibility

### Effect

Danger to personnel and property

Low operating efficiency

Eventually, high repair and replacement costs

Poor operation

### Prevention

Establish a definite maintenance policy and procedure

Assign maintenance responsibility
Insist on performance of maintenance functions
Keep maintenance log

### Operating Hazard
Condensate tank explosion

### Cause
Improperly vented tank
Vent too small
Vent is trapped
Frozen condensate in trapped vent

### Effect
Tank pressure may exceed design pressure
Tank may explode
Loss of life
Property damage

### Prevention
Eliminate traps in vent line
Eliminate restrictions in vent line
Vent to be full size—no valves
Vent to be run vertically from tank

## Furnace Explosion

### Operating Hazard
Inadequate pilot/ignitor

### Cause
Low gas pressure
Low fuel oil pressure
Improperly positioned pilot/ignitor

Nozzle too small

Plugged orifice

Improper light-off damper setting

### Effect

May not ignite the main flame

Delayed ignition

Fireside explosion

Fire

Boiler damage

Loss of life and/or injury to personnel

Property damage

### Prevention

Periodic pilot maintenance

Properly positioned pilot

Periodic pilot verification test

Use procedures illustrated in NFPA 85 Series

### Operating Hazard

Delayed ignition

### Cause

Inadequate pilot/ignitor

Low fuel oil pressure

Insufficient fuel rate

Excessive air rate

Low oil temperture

Water in fuel

### Effect

Fireside explosion

Fire

Boiler damage

Loss of life and/or injury to personnel
Property damage

### Prevention
Provide adequate pilot
Correct light-off fuel-air ratio setting
Avoid excessive restart attempts
Review or follow manufacturer's instructions
Conduct pilot turn down test

### Operating Hazard
Insufficient combustion air

### Cause
Lack of or insufficient boiler room air openings
Dirty combustion air blower
Combustion air blower running too slow or slipping
Incorrect fuel-air ratio setting
Blower inlet blockage
Outlet damper blockage
Plugged boiler gas passage

### Effect
Poor combustion
Delayed ignition
Fireside explosion
Loss of life and/or injury to personnel
Property damage
Fire
Boiler damage
Increased emissions

### Prevention
Provide adequate air to boiler room
Keep combustion air fans clean and running at the proper speed

Periodically observe: dampers, air inlets and outlets, combustion controls, boiler gas passages, hot flue gas temperature

### Operating Hazard

Tampering with combustion safety control

### Cause

Deliberate action by personnel
Lack of knowledge on the part of personnel
Inadequate operator training

### Effect

Fireside explosion
Fire
Loss of life and/or injury to personnel
Boiler damage
Property damage

### Prevention

Review and follow manufacturer's instructions
Prevent access by unauthorized personnel by locking equipment cabinets
Train operators in proper maintenance procedure

### Operating Hazard

Manual operation of combustion safety control (programmable)

### Cause

Deliberate action by personnel

### Effect

May cause ignition of main flame at the wrong time
Fireside explosion
Loss of life or injury to personnel
Boiler damage

Property damage

Fire

### *Prevention*

Do not operate combustion safety control manually

Review and follow the manufacturer's instructions

Provide adequate training for operators

Prevent access of unauthorized personnel by locking equipment cabinets

### *Operating Hazard*

Leaking fuel safety shutoff valves

### *Cause*

Defective valve

Foreign matter under valve seat

### *Effect*

Fuel flows to the boiler

Uncontrolled ignition of fuel

Fireside explosion

Loss of life and/or injury to personnel

Boiler damage

Property damage

Fire

## Implosions

### *Operating Hazard*

Excessive negative pressure in combustion zone

### *Cause*

Flame out

Induced-draft fan runaway

*Effect*

Equipment damage resulting in injury to personnel

*Prevention*

Maintain proper operation of control equipment

Do not bypass control equipment

Use the procedures suggested in NFPA Article 85 G

# REVIEW QUESTIONS

1. How often should boiler room employees receive safety training?
2. What type of clothing should be avoided when working around equipment?
3. What procedures must be followed when working on remotely started equipment?
4. What must new boiler room employees be warned about?
5. What test should be made before entering a confined or enclosed space?
6. What should be done when an unsafe practice and/or unsafe procedure is observed?
7. What should be done when airborne contaminants are found?
8. What should be done when moving parts are exposed?
9. What procedure should be followed when a fluid is spilled in the work area?
10. What should be done to prevent a fire in the boiler room?
11. Why should electrical box covers be closed at all times except when servicing is required?
12. How can an electrical box be opened without turning off the power?
13. Are electrical cheater plugs safe to use?
14. Why are superheated steam leaks very hazardous?
15. What could be the result of flue gas leaks?
16. What should be done before walking on fly ash?
17. When firing a boiler with fuel oil, what could cause oil gasification?

18. What should be done to steam lines to prevent wet steam when using a steam-atomizing fuel oil burner?

19. How could abrasive material in the fuel oil cause flareback?

20. Why would a gas relief valve discharge to the atmosphere?

21. What should be done for boilers using multiple gas sources?

22. What would cause a fuel rich gas-air mixture in a boiler furnace?

23. What may be the result of defective safety valves?

24. What could a defective low water cutoff on a boiler cause?

25. How is maximum safety achieved in the work place?

# CHAPTER 3

# TYPES OF BOILERS

## Learning Objectives

The objectives of this chapter are to:

- Introduce the theory of steam generation to the reader
- Acquaint the reader with the different types of boilers
- Make the reader aware of the purpose of the chimneys, draft fans, and breechings used in boiler installations
- Acquaint the reader with the different parts of a boiler.

A boiler is an enclosed vessel in which water is converted to steam by burning fuel and raising the pressure and temperature of the water and steam for the boiler's intended purpose.

A central boiler plant may have one or more boilers which utilize gas, oil, wood, or coal as the fuel. The steam generated is used to heat buildings and to provide hot water and steam for cleaning, sterilizing, cooking, laundry and industrial operations. Small heating boilers also provide steam and hot water for small buildings.

This chapter will help you acquire a useful knowledge of steam generation, types of boilers, parts of a boiler, and the various fittings commonly found on boilers. A main objective of this chapter is to lay the foundation so that skill in the operation, maintenance, and repair of boilers can be developed.

## STEAM GENERATION THEORY

Suppose that you set an open pan of water on the stove and turn on the heat. You will find that the heat causes the temperature of the water to increase and, at the same time, expand in volume. When the temperature reaches the *boiling point* (212°F, at sea level), a physical change occurs in the water: The water starts vaporizing. If you hold the temperature at the boiling point long enough, the water will continue to vaporize until the pan is dry. An important point to remember is that *the temperature of water will not increase beyond the boiling point.* Even if you add more heat after the water starts boiling, the water will not get any hotter as long as it remains at the same pressure.

But suppose that you place a close-fitting lid on the pan of boiling water. The lid prevents the escape of steam, and this results in a buildup of pressure inside the container. However, if an opening is made in the lid, steam will escape at the same rate as it is generated. As long as any water remains in the vessel, and as long as the pressure remains constant, the temperature of the water and steam will remain constant and equal.

The steam boiler operates on the same basic principle as a closed container of boiling water. It is true with the boiler as well as with the closed container that steam formed in boiling tends to push against the surface of the water and the sides of the vessel. Because of this downward pressure on the surface of the water, a temperature in excess of 212°F is required for boiling. The higher temperature is obtained simply by increasing the supply of heat. Bear in mind, therefore, that *an increase in pressure means an increase in the boiling point temperature.*

There are a number of technical terms used in connection with steam generation. Some of the commonly used terms that you should know are listed below.

*Degree:*   A degree is defined as a measure of heat intensity.

*Dry Saturated Steam:*   Dry saturated steam is steam at the saturation temperature corresponding to a given pressure; it contains no water in suspension.

*Heat:*   Technically speaking, heat is a form of energy which is measured in units known as British Thermal Units (Btus). One Btu is the amount of heat required to raise one pound of water one degree Fahrenheit at sea level.

*Latent Heat:*   Latent heat is hidden heat that cannot be felt or measured with a thermometer. It is the amount of heat required to change the state of a substance without changing its temperature.

*Quality:*   The quality of steam is expressed in terms of percent. For instance, if a quantity of wet steam consists of 90% steam and 10% moisture, the quality of the mixture is 90%.

*Sensible Heat:*   The term sensible heat refers to heat that can be measured with a thermometer.

*Steam:*   Simply stated the term steam means water in a vapor phase.

*Superheated Steam:*   Superheated steam is steam at a temperature higher than its saturation temperature corresponding to a given pressure. For example, a boiler may operate at 415 pounds per square inch gauge. The corresponding saturation temperature for this pressure is 445°F and this will be the temperature of the water in the boiler and the steam in the drum. (Charts and graphs are available from the ABMA, ASHRAE, and other such organizations for use in computing this pressure-temperature relationship.) This steam can be passed through a superheater where the pressure will remain about the same but the temperature will be increased to some higher figure.

*Temperature:*   Temperature may be defined as a measure in degrees of sensible heat.

*Wet Saturated Steam:*   Wet saturated steam is steam at the saturation temperature corresponding to a given pressure; it contains water particles in suspension.

You know, of course, that all the water in a vessel, if held at the boiling point long enough, will change into steam. What you may not know, is that there is a definite relationship between the pressure and the temperature. As long as the pressure is held constant, the temperature of the steam and boiling water will remain the same.

# BOILER DESIGN REQUIREMENTS

It is essential that a boiler meet certain requirements before it is considered satisfactory for operation. Three important requirements are that the boiler be (1) safe to operate, (2) able to generate steam at the desired rate and pressure, and (3) economical to operate.

Make it a point to familiarize yourself with the Boiler Code and other requirements applicable to the area in which you are located.

Here are a few rules set up by the American Society of Mechanical Engineers (ASME). These show the general guidelines used by engineers when designing the boilers that you will use.

For economy of operation, and to generate steam at the desired rate and pressure, a boiler must have

1. Adequate water and steam capacity
2. Rapid and positive water circulation
3. A large steam-generating surface
4. Heating surfaces which are easy to clean on both water and gas sides
5. Parts accessible for inspection
6. A correct amount and proper arrangement of heating surface
7. Firebox for efficient combustion of fuel

# TYPES OF BOILERS

There are two general types of boilers in operation today: the fire-tube boiler, and the water-tube boiler. Our discussion here will be concerned with the different features of these types of boilers. The main distinction between the two types is as follows:

1. Fire-tube boilers are those in which the products of combustion pass through and the water surrounds the tubes.

2. Water-tube boilers are those in which the products of combustion surround and the water flows through the tubes.

## Scotch Marine Fire-Tube Boilers

The Scotch Marine type of fire-tube boiler is especially suited to a great many applications. (See Figure 3–1.) They are available in portable-type units which can be moved with ease and with a minimum amount of foundation work. This type of boiler is a complete, self-contained unit and its design includes automatic controls, steel boiler, and burner equipment. These advantages are very important in temporary installations because no disassembly is required when the boiler needs to be moved to a more suitable location. Scotch Marine boilers are also manufactured in larger stationary types for use in permanent installations.

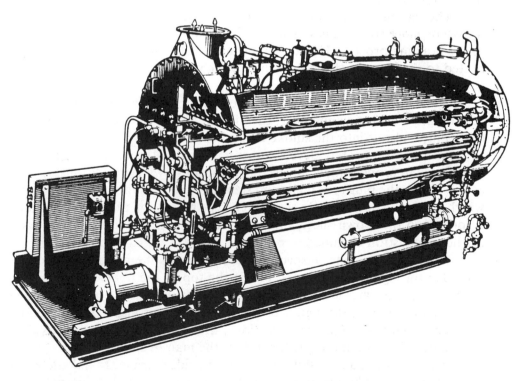

**FIGURE 3-1.**
Scotch Marine fire-tube boiler

The Scotch-type boiler has a two-pass (or more) arrangement of horizontal tubes. This allows the heat inside the tubes to travel back and forth before leaving the boiler. They are also equipped with an internally fired furnace, which has a cylindrical combustion chamber. There is a flue gas outlet, or smoke breeching, located on the front end of the boiler. Oil is the fuel commonly used to fire the Scotch-type boiler. Where desirable, though, it can be fired by gas, coal, or wood.

One advantage that the Scotch-type fire-tube boiler has over the water-tube boiler is that it requires less space, and can be set up in a low ceiling room. Also, the tubes are all the same size which saves time and trouble in making tube replacements.

The Scotch-type boiler also has a few disadvantages. Its shell runs from 6 to 8 feet in diameter, a type of construction which requires a large amount of reinforcement. Also, the fixed dimensions of its internal furnace causes some difficulty in cleaning the surfaces of the section below the combustion chamber. Another drawback is the limited capacity and pressure of the Scotch-type boiler.

The setting of the Scotch-type boiler is self-supporting. The shell rests in two or more steel cradles. The boiler is sometimes pitched slightly to aid in the draining process. The setting includes a blowdown pipe which is connected to the bottom of the steel shell, which in turn, is screwed into a pad that is riveted to the shell.

The flow of gases in a two-pass Scotch-type boiler is first toward the rear of the combustion chamber; then it is returned by way of the tubes to the front, and out into the smoke box and stack breeching.

An important safety device sometimes used is the fusible plug, which is designed to provide added protection against low water conditions. In case of low water, the plug core melts and steam escapes, resulting in a loud noise which warns the operator. On the Scotch-type boiler, the fusible plug usually is located in the crown sheet, but sometimes it is found in the upper back of the combustion chamber.

Access for cleaning, inspection, and repair of the boiler watersides is provided through two manholes: one in the top of the boiler shell, and the other one in the water leg. The *manhole* is an opening which is large enough that a man can enter through it to the boiler shell for inspection, cleaning, and repair purposes. On such occasions, always make sure that all valves are secured, locked, and tagged, and that the man in charge knows you are in there. Also, have a man stationed at the outside entrance to aid and assist in the operation. The *handhole* is an opening large enough to permit hand entry for cleaning, inspection, and repairs to headers and tubes.

## Horizontal Return Tubular Boiler

These are considered to be more of the stationary-type boilers than some of the Scotch Marine boilers. A stationary boiler can be defined as one having a permanent foundation and not easily moved or relocated. A popular type of stationary fire-tube boiler is the horizontal return tubular (HRT) boiler. (See Figure 3–2.)

The initial cost of an HRT boiler is relatively low. Also, their installation is not too difficult. The boiler setting can be readily changed to meet the different fuel requirements of coal, oil, wood, or gas. Tube replacement is also relatively easy since all tubes in the HRT boiler are the same size, length, and diameter.

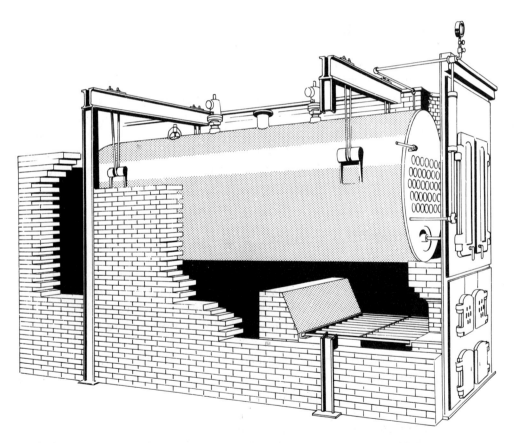

**FIGURE 3-2.**
Horizontal return tubular (HRT) fire-tube boiler

The gas flow in the HRT boiler is from the firebox to the rear of the boiler, then it returns through the tubes to the front where it is discharged to the breeching and stack.

The HRT boiler has a pitch of 1 to 2 inches to the rear. This allows sediment to settle toward the rear near the bottom blowdown connection. The fusible plug is located 2 inches (50.80 mm) above the top row of tubes. Boilers over 40 inches (1016 mm) in diameter require a manhole in the upper part of the shell. Those over 48 inches in diameter must have a manhole in the lower, as well as in the upper part of the shell. Do not fail to familiarize yourself with the location of these and other essential parts of the HRT boiler. The knowledge you acquire will be a great help to you in the performance of duties involving boilers.

## Firebox Boilers

Another type of fire-tube boiler is the firebox boiler, which generally is used for stationary purposes. (See Figure 3–3.) Firebox boilers require no setting except possibly an ash pit (when using coal as the fuel). As a result, they can be quickly installed and placed in service. The gases in a firebox boiler make two passes through the tubes. They travel from the firebox through a group of tubes to a reversing chamber, and return through a second set of tubes to the flue connection on front of the boiler, and are then discharged up the stack.

## Water-Tube Boilers

Water-tube boilers may be classified in a number of ways. For our purposes, though, let us classify them as straight-tube and bent-tube. These two classes will be discussed separately in succeeding sections. To avoid confusion, make sure you study each illustration referred to throughout the discussion.

The straight-tube category of water-tube boilers includes three types: (1) sectional-header cross drum, (2) box-header cross drum, and (3) box-header longitudinal drum.

**Sectional-Header Cross Drum.**   The headers of the sectional-header cross drum boiler are steel boxes into which the tubes are rolled. (See Figure 3–4.) The feed water enters and passes down through the downcomers (pipes) into the rear sectional headers from which the tubes are supplied. The water

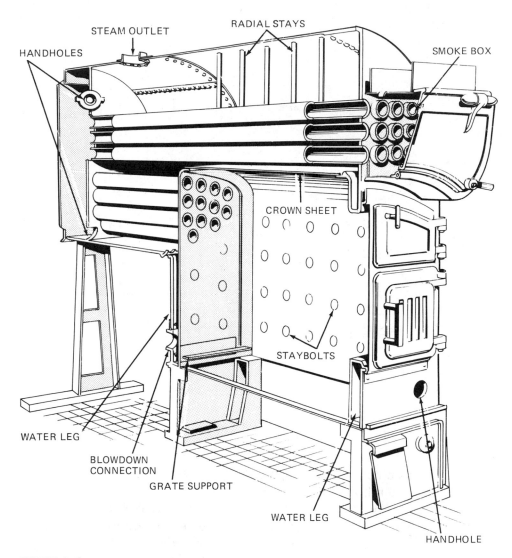

**FIGURE 3-3.**
Split section of small firebox boiler

is heated and some of it changes into steam as it flows through the tubes to the front headers. The steam-water mixture returns to the steam drum through the circulating tubes and is discharged in front of the steam-drum baffle which helps to separate the water and steam.

The steam is removed from the top of the drum through thc dry pipe.

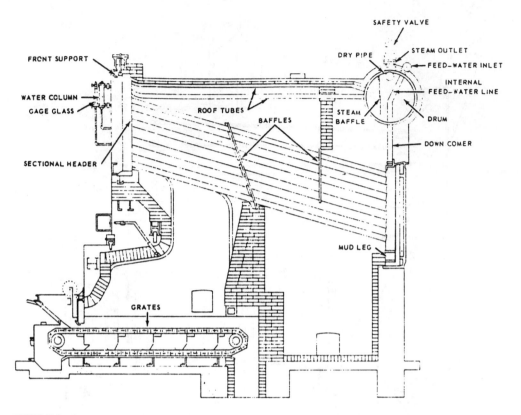

**FIGURE 3-4.**
Sectional-header cross drum boiler with chain grate stoker

This pipe extends along the length of the drum. The steam enters this pipe through holes or slots in the top half.

The headers, the distinguishing feature of this type of boiler, are usually of forged steel and are connected to the drums with tubes. The headers may be vertical, as in Figure 3-4, or at right angles to the tubes. The tubes are rolled and flared into the header. A handhole is located opposite the ends of each tube; this facilitates the inspection and cleaning process. A mud drum is connected to the bottom of each header by short nipples. Its purpose is to collect sediment, which is removed by blowing down the boiler.

Baffles are usually arranged so that gases are directed across the tubes three times before being discharged from the boiler below the drum.

**Box-Header Cross Drum Boiler.**    In the box-header cross drum boiler, the box headers are shallow boxes made of two plates: (1) a tube-sheet plate, which is bent to form the sides of the box and (2) a plate containing the handholes, which is riveted to the tube-sheet plate. (See Figure 3-5.)

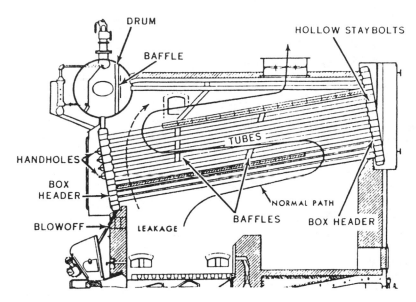

**FIGURE 3-5.**
Box-header cross drum boiler

Some box-header cross drum boilers are designed so that the front plate can be removed for access to the tubes. The tubes enter at right angles to the box header and are expanded and flared in the same manner as the sectional-header boiler. The boiler is usually built with the drum in front. It is supported by lugs fastened to the box headers. This boiler has either cross or longitudinal baffling, arranged to divide the boiler into three passes.

The water enters the bottom of the drum, then flows through the connecting tubes to the box header, and then through the tubes to the rear box header, and back to the drum.

**Box-Header Longitudinal Drum Boiler.** These types of boilers may have either a horizontal or an inclined drum. The box headers are fastened directly to the drum when the drum is inclined. If the drum is horizontal, the front box header is connected to it at an angle greater than 90 degrees. The rear box header is connected to the drum by tubes. Longitudinal or cross baffles can be used with either type.

**Bent-Tube Boilers.** There are many types of bent-tube boilers. Boilers of this type usually have three drums. (See Figure 3–6.) The drums are usually of the same diameter and are positioned at different levels with each other. The uppermost or highest positioned drum is referred to as the *steam drum,* the center positioned drum is referred to as the *water drum,* and the lowest

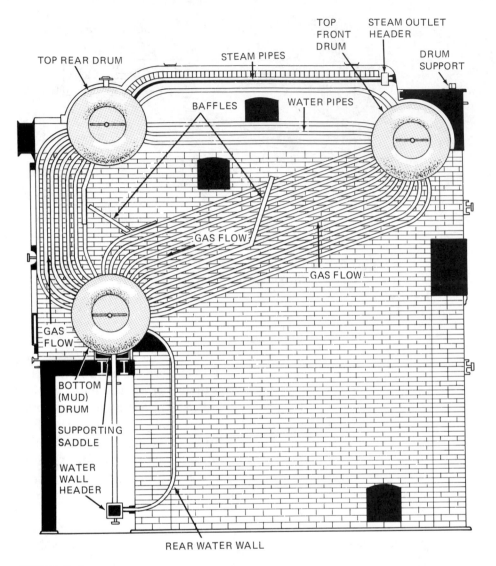

**FIGURE 3-6.**
Three-drum bent-tube boiler with longitudinal baffle arrangement

drum is the *mud drum*. The drums are connected by tube banks. The tubes are bent at the ends to enter the drums radially.

Water enters the top rear drum, passes through the tubes to the bottom drum, and then moves up through the tubes to the top front drum. A mixture of steam and water is discharged into this drum where the water is separated from the steam. This is accomplished by the steam rising to the top of the drum. The water is returned to the top rear drum through the water circulators. At this point, the rest of the steam is separated from the water by a baffle and/or dry pipe. Notice that the steam pipe extends from the top front drum to the top rear drum. Its purpose is to carry steam to the top rear drum. The main steam outlet is connected to the top rear drum.

Notice that the waterwall is fed from a waterwall header which is not exposed to the furnace heat. Water circulates upward in the waterwall tubes to the bottom drum, and downward from the drum to the waterwall heater.

**Four-Drum Stirling Boiler.**    This boiler has three upper steam and water drums and one lower mud drum. (See Figure 3–7.) The upper drums are

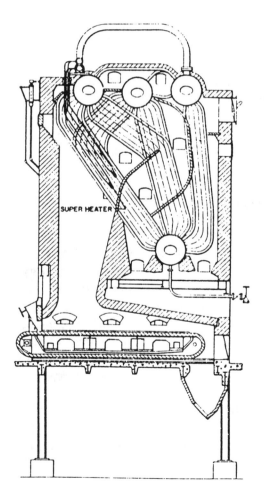

**FIGURE 3-7.**
Four-drum stirling boiler

interconnected by steam or water circulator tubes. These boilers have good water circulation because they are quite high and have ample hydraulic head. The illustrated boiler has a superheater. The baffle arrangement produces a cross gasflow.

Many types of baffle arrangements are used with the bent-tube type boilers. Usually, they are installed so that 70 to 80 percent of the heat will be absorbed by the inclined tubes between the lower drum and the top front drum.

The water-tube boilers discussed above offer a number of worthwhile advantages. For one thing, they afford flexibility in starting up. They also have a high productive capacity, which ranges from 100,000 to 1,000,000 pounds of steam per hour. In case of tube failure, there is little danger of a disastrous explosion with the water-tube type of boiler. The furnace can not only carry a high overload, but it can also be easily modified for firing by either oil or coal. Still another advantage is that there is minimal difficulty in getting to sections inside the furnace for cleaning and repair purposes.

There are several disadvantages that are common to water-tube boilers. There is a high construction cost involved, a factor which poses one of the main drawbacks of this type of boiler. The large assortment of tubes required of this boiler, and the excessive weight per unit weight of steam generated, are other unfavorable factors.

## BOILER PARTS AND FITTINGS

By now you should have a general idea of the overall basic structure of a boiler. A number of questions probably have come to mind as to the importance of certain boiler parts and the operation or function of various devices, such as parts and fittings. For any unit or device not covered in this text, check the manufacturer's manual for information on details of its construction and method of operation (where applicable).

The term *fittings* includes various controlling devices on the boiler. Bear in mind, therefore, that fittings are vitally important to the economy of operation and safety of the personnel and equipment. A thorough knowledge of fittings is necessary if you are to acquire skill in the installation, operation, and servicing of steam boilers. Refer to chapters 6, 7, 9, 11, and 12 in this text for the operation of the various types of controlling devices used on boilers.

## Settings

The setting is the structure that encloses the boiler and forms the furnace. It may also support all or part of the boiler. The self-supported type of setting is common. The outer wall is built of hard-burned brick. The inner wall is built of firebrick to withstand high temperatures. Several examples of this type of setting are shown in Figure 3–8.

## Chimneys, Draft Fans and Breechings

Chimneys are necessary for discharging the products of combustion at an elevation high enough to comply with health requirements, and to prevent a nuisance due to low flying smoke, soot, and ash. A boiler needs draft to mix air correctly with the fuel supply, and to conduct the flue gases through the complete setting.

The air necessary for combustion of fuel cannot normally be supplied by natural draft. Therefore, draft fans may be used to ensure that the air requirements are properly met. Two types of draft fans used on boilers are forced-draft and induced-draft fans. They are damper-controlled and usually are driven by an electric motor.

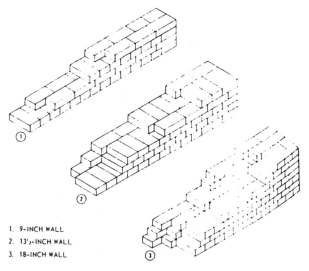

1. 9-INCH WALL
2. 13½-INCH WALL
3. 18-INCH WALL

**FIGURE 3-8.**
Furnace sidewall construction

**Forced-Draft Fan.**    The forced-draft fan forces air through the fuel bed, or fuel oil burner, and into the boiler furnace to supply air for combustion of the fuel.

**Induced-Draft Fan.**    The induced-draft fan draws gases through the setting, thus facilitating their removal through the stack.

**Breechings.**    Breechings are used to connect the boiler to the stack. They are usually made of sheet steel, with provision for expansion and contraction. The breeching may be carried over the boilers, in back of the setting, or even under the boiler room floor. Keep breechings as short as possible, and free from sharp bends and abrupt changes in area. The cross-sectional area should be approximately 20 percent greater than that of the stack, to keep the draft loss to a minimum. A breeching with circular cross section causes less draft loss than one with rectangular or square cross section.

## Major Construction Features

A boiler must meet certain requirements before it is considered to be satisfactorily installed. It must be safe to operate, it must generate clean steam at the desired rate and pressure, and it must be economical to operate. A set of rules for the construction of stationary steam boilers, known as the American Society of Mechanical Engineers Boiler Construction Code, has been widely adopted by insurance underwriters and governmental agencies. This code contains mandatory provisions for methods of construction and installation, materials to be used, design features, and recommended operating and inspection procedures. Standard practice requires that a qualified boiler inspector make an inspection at least once a year.

**Stays and Braces.**    The flat surfaces of a boiler are held in place by braces, which are called stays. The most important types of stays are the gusset, diagonal, through, slingstay, and staybolt. These are characterized as follows:

1. The gusset stay consists of one or two angles fastened to each of the surfaces and connected by a steel plate. (See Figure 3–9). They are very rigid and sometimes cause undue stress.
2. Diagnoal stays are normally made of round iron flattened on one end and T-shaped on the other. (See Figure 3–10.) The T-shaped end is

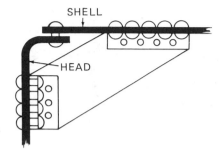

**FIGURE 3-9.**
Gusset stay

riveted to the flat surface to be supported and the other end is riveted to the boiler shell.

3. Through stays extend from one flat surface to another. One type is threaded and fitted with two nuts and a large washer on each end. Another type is threaded on one end; the other end is supported by a pin connection to two angles riveted to the head. (See Figure 3–11.) To permit water circulation between the head and angles, they are separated with spacers.

4. Slingstays are similar to the radial stays. They have a pin-type joint at one or both ends to permit a slight movement of the stayed sheets. (See Figure 3–12.) Figure 3–13 shows the slingstay bracing of a firebox-type boiler.

5. Staybolts are short through stays which are threaded through the tube sheets and riveted at the ends. A 3/16-inch diameter telltale hole is drilled in each end of the staybolt; leakage from the hole indicates a broken staybolt. The holes may be drilled to at least 1/2 inch beyond the inside of the plate, or all the way through the bolt. (See Figure 3–14.)

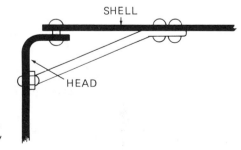

**FIGURE 3-10.**
Diagonal stay

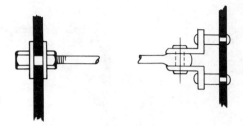

**FIGURE 3-11.**
Through stay, showing two methods of
fastening to shell

If the staybolt is over 8 inches, it need not be drilled. The flexible stay, a variation of the staybolt, usually has a ball-and-socket connection on one end and is screwed into the sheet and riveted over on the other end. Some types are screwed through the sheet and fitted with a special cap.

**Internal Components.** The boiler, for construction purposes, can be divided into two primary components: the furnace, which is designed to give efficient fuel combustion; and the pressure parts, which are used to enclose the steam or water.

*Furnace.* Proper furnace design is an important factor in obtaining the efficient combustion of the fuel. Its volume, that is, the combustion space, determines to some extent the time available for complete combustion. The shape of the furnace can also help promote turbulence. Some furnaces have refractory arches and/or bridges that reflect heat and maintain a high temperature in specific zones. These are required to burn some kinds of fuels. The heat release in Btu per cubic foot of furnace volume must be kept within economical limits. These limits depend on the nature of the furnace setting (enclosing walls), boiler heating surface exposed to the radiant heat, nature of the fuel, and type of firing.

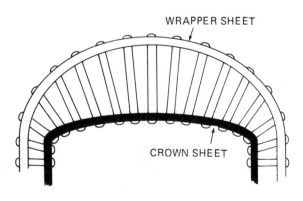

WRAPPER SHEET

CROWN SHEET

**FIGURE 3-12.**
Radial stays supporting crown
sheet

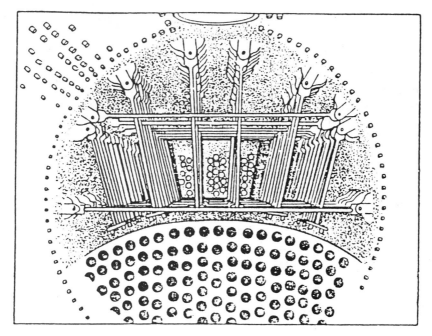

**FIGURE 3-13.**
Inside of 300 hp 150 psi firebox boiler, showing slingstay bracing of crown and wrapper sheet

The furnace enclosure or setting may consist of all solid refractory walls, refractory air-cooled walls, a combination of refractory and water-cooled walls, and waterwalls when the boiler heating surface forms the entire furnace wall.

Excessive temperature adversely affects refractory materials; therefore, refractory wall furnaces are limited to operations in which heat is released. Permissible heat release of a furnace may range from 20,000 to 150,000 Btu per cubic foot. Usually furnace installations are designed for a heat release of 25,000 to 35,000 Btu per cubic foot, with a peak load of 50,000 Btu.

***Pressure Parts.***    Pressure parts are the totally enclosed metallic sections, compartments, or tubes which contain steam or water. They must be strong enough to continuously withstand the maximum pressure and tem-

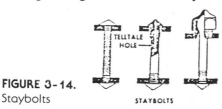

**FIGURE 3-14.**
Staybolts

perature for which the boiler is designed. These sections comprise the drums, headers, waterboxes, tubes, and waterwalls.

Tubes and drums are used as the principal pressure parts largely because of their ease of manufacture and because a cylinder, as such, has maximum strength for minimum material thickness. Tubes can be bent readily and arranged in groups for effective heat absorption at minimal cost.

***Headers.***     Headers are used in waterwall construction to connect groups of tubes. They are usually formed of heavy wall tubing or pipe, but may be forged. (See Figure 3–15.)

For operating pressures up to about 500 psig, tubes are joined to drums and headers by rolling. The tubes and flues are installed by expanding them

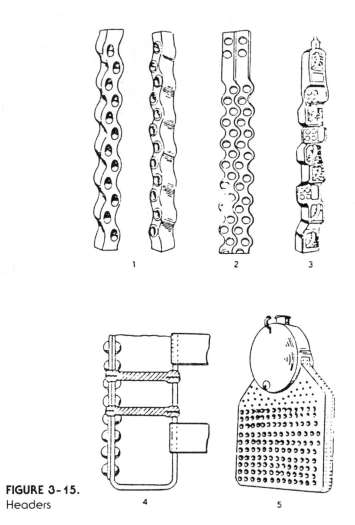

**FIGURE 3-15.**
Headers

into seats which may or may not be grooved. The expander tools consist of a tapered pin which fits in a cage containing several small rollers. A different size expander is required for every size tube or flue.

Boiler tubes are installed with the ends projecting 1/4 inch, to a maximum of 3/8 inch, beyond the tube sheets. Projecting ends are flared slightly in water-tube boilers and are left that way because they are surrounded by water or steam. Since tube ends of a fire-tube boiler are surrounded by hot gases, they would soon burn off if allowed to project. Therefore, they are beaded and hammered until they are flat against the tube sheet. The process also increases the holding power of the tube.

Tube holes must be drilled full size. In high-pressure boilers with heavy drum walls, tube holes are often counterbored to give the tube a narrow seat which is easier to make and keep tight.

*Baffles.* Baffles are thin walls or partitions used to direct the combustion gases over the heating surfaces with minimal draft loss and maximum heat absorption. Baffles should be designed and installed to conform with the reduction in gas volume which results when heat absorption lowers the gas temperature. When they are designed this way, constant gasflow velocities with minimal draft loss are obtained. Baffles can be designed to produce a cross flow of gas (see Figure 3–4), or to cause a longitudinal flow (see Figure 3–5). Baffles may be constructed of either refractory material or metal. Steel baffles are used in low temperature zones; refractory baffles, supported and cooled by the tubes, are used in high temperature zones.

*Downcomers.* Downcomers are the large tubes that are located outside the furnace in which the flow of water is always downward. In the single pass boiler, two downcomers connect the steam drum to each front side wall header. One of these downcomers enters at the top and the other enters in the middle of the header. The lower rear header is also connected by two downcomers from the steam drum to the header, entering the header at each end.

*Risers.* Risers are smaller tubes in which the flow of water and steam is always upward. The single pass boiler has two risers for each sidewall header; one connected to the top of the header, and the other connected to the middle. Both risers discharge to the steam drum. The upper end of each tube in the rear wall is expanded into its corresponding sinuous header alone.

*Dry Pipe.* The dry pipe is a large pipe enclosed at both ends and secured in the top of the steam space. The top half of this pipe is slotted or perforated. The main and auxiliary steam stops are connected to the dry pipe for obtaining dry steam. The dry pipe is used to prevent water turbulence within the boiler and to prevent moisture carryover into the steam distribution system.

*Internal Feedlines.*    An internal feedline is a continuation of the feed-water line located below the water level in the steam drum. It is perforated and runs the length of the steam drum. Its purpose is to provide even distribution of water so that there is no stress or strain in one area of the drum.

# REVIEW QUESTIONS

1.  What effect does an increase in steam pressure inside a boiler have on the boiling point of the water inside?
2.  What is steam that contains no water in suspension known as?
3.  In what terms is the quality of steam expressed?
4.  What is superheated steam?
5.  What will cause the temperature of the steam and boiling water in a boiler to remain the same?
6.  With what should you familiarize yourself to know the boiler requirements in your area?
7.  Name the two general types of boilers in operation today.
8.  What type of boiler is the Scotch Marine boiler?
9.  Why are boilers sometimes installed with a pitch to the rear?
10.  What is a fusible plug?
11.  Define a stationary boiler.
12.  Why are the tubes in an HRT boiler easy to replace?
13.  When is an ash pit used with a firebox boiler?
14.  Name the three types of straight-tube water-tube boilers.
15.  What is the distinguishing feature of the sectional-header cross drum boiler?
16.  What is the purpose of the baffles used in boilers?
17.  What is the uppermost drum of a bent-tube boiler known as?
18.  Why are the tubes in a bent-tube boiler bent?
19.  How many mud drums are there in a four-drum stirling boiler?
20.  Define a boiler setting.

21. What part of the boiler is used to transfer the products of combustion outdoors?
22. What is the purpose of boiler draft?
23. Why are fans used to produce boiler draft?
24. Name the three types of boiler draft systems.
25. What code contains mandatory provisions for installing and operating boilers?
26. With what are the flat surfaces of a boiler held in place?
27. What is the purpose of the telltale hole in staybolts?
28. Why is the shape of a boiler furnace important?
29. Why are refractory wall furnaces limited to operations in which heat is released?
30. What is the purpose of the dry pipe in a boiler?

# CHAPTER 4

# TEMPERATURE AND THE EFFECTS OF HEAT

## Learning Objectives

The objectives of this chapter are to:

- Make the reader aware of the differences between the Fahrenheit and the Celsius (centigrade) temperature scales
- Acquaint the reader with the different types of heat
- Show the reader the different methods of heat transfer
- Acquaint the reader with the different effects of heat.

In everyday life, either consciously or unconsciously, we experience temperature and heat and the effects that they have on our surroundings. Boiler technicians and operators should be more aware of these facts than anyone else. It is their responsibility to maintain equipment that depends upon heat and heat transfer.

# TEMPERATURE

Temperature is defined as the measure of degree or intensity of heat present in a substance. Temperature is not a measure of the amount of heat present, just the intensity of it. The degree of heat is called temperature and the instrument for measuring temperature is known as the thermometer.

## Thermometer

This device takes advantage of one of the most important effects of temperature, that is, *expansion* or the increase in size of a body when its temperature is raised. A thermometer consists of a glass tube containing a fluid, usually mercury or colored alcohol. This fluid only partly fills the tube; the space above it is under a vacuum. There is an enlarged reservoir at the base of the tube to hold a supply of the fluid. A rise in temperature causes the fluid to expand in the bulb and then rise in the tube, which is marked off in a graduated scale. There are several different scales which are used for this purpose. The Fahrenheit scale and the Centigrade (Celsius) scale are the most common scales.

**Fahrenheit Scale.**   The Fahrenheit scale is the scale used in the United States for the most part, except in laboratory experiments. (See Figure 4–1.) It was first invented by a man named Fahrenheit in about 1724. To locate the graduations on the scale, the thermometer is placed in a mixture of pure ice and water and the fluid column drops to a certain point. This point is marked 32°F, the freezing and melting point of ice. Then the tube is immersed in boiling water and the fluid rises to a high point, which is marked 212°F. This allows a difference of 180 degrees, or divisions, between these two points on the Fahrenheit scale. The scale is equally divided into 180 spaces between the two points. These measurements must be taken at sea level because an atmospheric pressure lower than sea level changes the location of the boiling and freezing points. This scale extends down to absolute zero which is

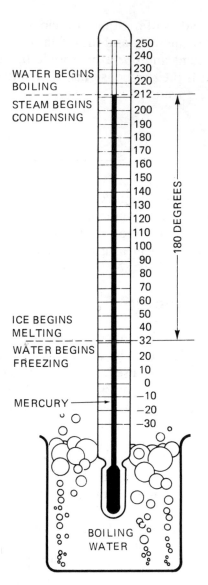

**FIGURE 4-1.**
Fahrenheit thermometer scale

−460°F and upward to indicate the amount of temperature present in the substance being measured.

**Celsius (Centigrade Scale).**    On this scale, the freezing point of water is marked at 0°C and the boiling point is marked at 100°C. (See Figure 4-2.) Absolute zero on this scale is indicated to be −273°C. Notice that there are 100 divisions between the freezing and boiling points on this scale.

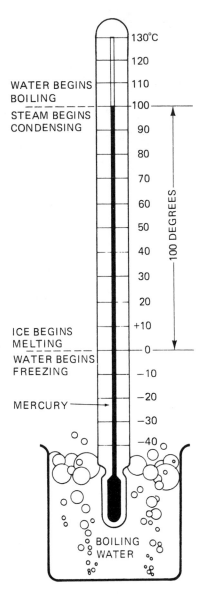

**FIGURE 4-2.**
Celsius (centigrade) temperature scale

Temperatures below zero on both thermometer scales are considered to be negative and are written thus: −29° (minus 29°). This figure is read as "29 degrees below zero Fahrenheit" or "Centigrade" as the case may be.

There is a great difference between the degrees indicated on the different thermometer scales. Because it is sometimes necessary to convert from one scale to another there are formulas especially for this purpose.

Formula 1: Fahrenheit Temperature $= 32 + (C° \times 1.8)$

Example

Change 60°C to Fahrenheit.

$$
\begin{aligned}
\text{Fahrenheit temperature:} \ &= 32 + (C° \times 1.8) \\
&= 32 + (60 \times 1.8) \\
&= 32 + 108 \\
&= 140°F
\end{aligned}
$$

Formula 2: Centigrade Temperature $= (F° - 32°)\, 1.8$

Example

Change 60°F to Centigrade.

$$
\begin{aligned}
\text{Centigrade temperature} \ &= (60 - 32)\, 1.8 \\
&= \frac{28}{1.8} \\
&= 15.55°C
\end{aligned}
$$

## Saturation Temperature

The saturation temperature of a fluid involves both the temperature and the pressure at which both a liquid and a vapor can exist in the same container at the same time. A liquid or gas is said to be at its saturation temperature when it is at its boiling temperature which corresponds to a given pressure. When there is an increase in the pressure over a fluid, the saturation temperature also increases. If there is a drop in the pressure over the fluid the saturation temperature also drops.

# HEAT

There is a difference between the meaning of the terms *heat* and *temperature*. Heat is usually considered the measure of quantity, while temperature is a measure of degree or intensity of heat. These two terms should not be confused. Two bodies may have the same temperature, yet due to their volume and their heat absorbing capacity, their heat content may be entirely different. For example, a quart of water and a gallon of water may be at the same temperature, but the gallon will contain four times as much as heat as the quart.

## Sensible Heat

Sensible heat is the heat which must be added to or removed from a substance to change its temperature but does not change its state. The word sensible means that the heat can be sensed with a thermometer or felt with the hand. An example of sensible heat may be seen when the temperature of an object is raised from 50°F to 75°F. (See Figure 4–3.) In this example, the change in temperature from 50 to 75°F is equal to 25°F which is sensible heat and can be measured with a thermometer.

**Specific Heat.**    Specific heat is the amount of heat required to change the temperature of 1 lb of a substance 1°F. This is sensible heat which can be measured with a thermometer and it is different for every different type of substance.

**Superheat.**    Superheat is the amount of heat added to a substance having a temperature above its boiling point. This is sensible heat and can be measured with a thermometer. If we heat water to its boiling point of 212°F at atmospheric pressure, then heat this steam to 230°F, the steam has a superheat of 18°F.

## Heat Measurement

Heat is measured in British Thermal Units which is abbreviated with the letters Btu. A Btu is defined as the amount of heat necessary to raise the temperature of one pound of pure water one degree Fahrenheit.

This unit measures the quantity of heat and has nothing to do with the

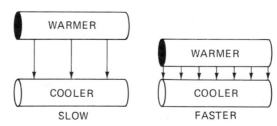

**FIGURE 4-3.**
Example of sensible heat

intensity, which is measured by degrees. The Btu indicates the amount of heat in a substance while degrees give the intensity or strength.

## Methods of Heat Transfer

The transfer of heat in boilers is very important for efficient operation of the equipment. Heat always flows from a warmer to a colder object. The greater the temperature difference, the faster the heat will flow. It is similiar to a car that will coast down a steep hill faster than it will coast down a small grade. Also, if two objects are in contact with each other, the transfer of heat will be faster than if they were separated. (See Figure 4-4.)

The three methods of heat transfer are (1) conduction, (2) convection, and (3) radiation.

**Conduction.**   Conduction is the transfer of heat through an object. The rate of heat transfer by this method depends upon the heat conductivity of the material. For example, a piece of glass will transfer heat much more slowly than will a piece of iron the same size. Heat transfer by conduction may take place between two objects that are in contact with each other. (See Figure 4-5.) For efficient heat transfer to occur the two surfaces must be in extremely close contact and they must be as clean as possible.

**Convection.**   A fluid is used in the transfer of heat by convection. This fluid may be either a gas or a liquid. In boilers this fluid is usually water. The heated fluids are lighter than the unheated fluids and will rise upward, while the heavier, cooler fluids will fall. In this manner a continuous flow is created by gravity. Heat transfer by convection occurs from the products of combustion to the boiler tubes which are carrying the water to be heated.

**Radiation.**   Heat transfer by radiation requires a high temperature difference and high temperatures to be efficient. Heat transfer by radiation is done through the use of wave motion. Heat transfer by radiation does not heat the air between the objects. In boilers, radiant heat is absorbed from the surrounding bricks and other equipment in the firebox. In many cases this is an important factor in boiler efficiency.

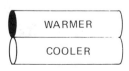

**FIGURE 4-4.**
Transfer of heat between objects

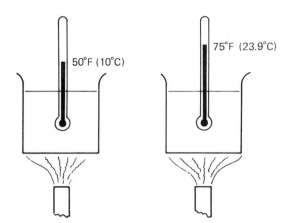

**FIGURE 4-5.**
Transfer of heat by conduction

During the operation of a boiler, heat will be transferred in all three of these ways at the same time. In some instances one will be greater than the others, but they all have an effect on the overall performance of the boiler and its related equipment.

## Latent Heat

Latent heat is hidden heat because it cannot be felt or sensed with a thermometer. It is the heat which is added to or removed from a substance which results in a change of state but no change in temperature, for example, ice to water, water to ice, water to steam or steam to water. There are four different types of latent heat: (1) latent heat of fusion, (2) latent heat of condensation, (3) latent heat of vaporization, and (4) latent heat of sublimation.

**Latent Heat of Fusion.**    The latent heat of fusion has a twofold purpose. First, it represents the amount of heat required to change a solid to a liquid. Second, it represents the amount of heat required to change a liquid to a solid. Both of these processes must take place at a constant temperature.

**Latent Heat of Condensation.**    The latent heat of condensation represents the amount of heat required to change a vapor to a liquid. This process must take place at a constant temperature.

**Latent Heat of Vaporization.**    The latent heat of vaporization represents the amount of heat required to change a liquid to a solid. Vaporization of water takes place at the surface of the liquid. This process must take place at a constant temperature.

*Ebullition.* Ebullition is a second part of evaporation. It is generally refered to as boiling and takes place in the liquid itself. The vapors (bubbles) rise through the liquid to its surface where they escape and become a vapor. There are three factors which govern the evaporation of a liquid.

1. Evaporization increases with an increase in temperature of the liquid.
2. Evaporization increases when the vapors are escaping into dry surroundings and slows down when they are escaping into wet surroundings.
3. Evaporization is increased with a decrease in the pressure exerted on the surface of the evaporating liquid.

**Latent Heat of Sublimation.** The latent heat of sublimation represents the amount of heat required to change a solid to a vapor with no evidence of it having gone through the liquid state. Not all substances are capable of going through this process.

## Effects of Heat

When any type of substance is heated it will expand a certain amount for each degree of temperature rise. Likewise, when any substance is cooled it will decrease in size (contract) a certain amount for each degree of temperature decrease. Water is the only exception to this rule. When water at a temperature above 39°F is cooled it will contract until a temperature of 39°F is reached. At this temperature it begins to expand a certain amount for each degree it is cooled. This is the reason that pipes burst when water is frozen. This expansion and contraction of the material causes a tremendous stress on the molecules and they will eventually break down and the material will be ruined.

There are three different types of expansion and contraction: (1) linear, (2) areal, and (3) volumetric.

**Linear Expansion.** Linear expansion is the amount of increase or decrease in the length of a piece of substance when it is either heated or cooled. It is known as the coefficient of linear expansion.

**Areal Expansion.** Areal expansion is the increase or decrease in the surface area of a substance when heated or cooled. It is generally taken as being two times the coefficient of linear expansion. This figure is sufficient for practical application of the rule.

**Volumetric Expansion.**   Volumetric expansion is the increase or decrease in the volume of a substance when heated or cooled. This type of expansion and contraction is important in the operation of boilers because all liquids, including water, increase in volume when heated. For all practical purposes, liquids are incompressible. If a hot water heating system is filled with cold water and then heated above 30°F the water will expand and the resulting pressure will become very great. This is the purpose of installing expansion tanks and safety relief valves in hot-water and steam-heating systems. The coefficients of volumetric expansion of liquids are difficult to determine because of the different types of materials used in the manufacture of the boiler and the circulating system.

# REVIEW QUESTIONS

1. When is a liquid or a gas at its saturation temperature?
2. What is heat?
3. What is temperature?
4. What is the amount of heat required to change the temperature of one pound of a substance 1°F?
5. Name the three methods of heat transfer.
6. Is radiant heat important in boiler operation?
7. What type of heat is required to change the temperature of a substance but does not change its state?
8. What type of heat cannot be felt or sensed with a thermometer?
9. What is the second part of the evaporation process?
10. Name the three different types of expansion and contraction.

# CHAPTER 5

# COMBUSTION THEORY

## Learning Objectives

The objectives of this chapter are to:

- Familiarize the reader with the theories of combustion
- Cause the reader to become aware of the different types of combustion
- Introduce the reader to the different fuel classifications
- Provide the reader with descriptions of the more practical combustion considerations regarding the equipment.

When a fuel (either solid, liquid or gas) is burned, there are certain phenomena which occur. These phenomena must be understood by the boiler operator for safety as well as for efficient operation of the equipment. He must understand the different requirements for each type of fuel commonly used in the production of hot water and/or steam and what the results will be if the basics of combustion are not followed.

# FUNDAMENTALS OF COMBUSTION

Combustion is the chemical process which takes place when a substance, such as the fuel for a boiler, is properly combined with oxygen and the process results in the release of heat and some light. There is a rapid transformation from the fuel to the heat produced by the combustion process. This process is sometimes referred to as oxidation. In most installations the oxygen for combustion is taken from the atmospheric air.

The conventional hydrocarbon fuels such as fuel oil, natural and LP gas, and coal are made up primarily of hydrogen and carbon in various compounds. When the combustion process is complete they produce mostly carbon and water vapor combined with small quantities of carbon monoxide. Small amounts of sulfur, oxidized to sulfur dioxide ($SO_2$) or sulfur trioxide ($SO_3$) during the combustion process, along with other noncombustible products such as ash, water and inert gases are also released.

## Basics of Combustion

The three basic requirements for complete combustion are (1) a temperature high enough to ignite the fuel, (2) oxygen, and (3) a fuel. These three requirements are shown with the combustion triangle in Figure 5–1.

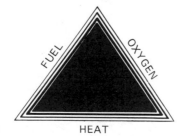

**FIGURE 5–1.**
Combustion triangle

When the fuel-air mixture is introduced into the furnace of the boiler, a pilot light or some other means of ignition increases the temperature of the mixture to its ignition temperature. The temperature of the mixture is then kept above the ignition temperature by the flame of the burning fuel. Should the temperature of the fuel-air mixture drop below the ignition temperature, the flame will be extinguished automatically. This condition must be guarded against. For example, the ignition temperature of natural gas is approximately 1,100°F. If the temperature of the mixture drops below this temperature the flame will go out and the gas-air mixture will enter the furnace unlighted. This could present a very hazardous situation if the safety devices should fail.

## Products of Combustion

Each type of fuel produces its own set of combustion products. Each product is different and requires its own burner and air adjustments. As an example, let us look at the products of combustion for natural gas. When 1 ft³ of natural gas is completely burned there are 8 ft³ of nitrogen, 1 ft³ of carbon dioxide, and 2 ft³ of water vapor produced, all of which are vented to the atmosphere. (See Figure 5–2.) These products in their natural form are not harmful to human beings.

However, the by-products of combustion are very hazardous to human beings. These by-products consist of carbon monoxide, aldehydes, keytones, oxygen acids, glycols, and phenols. Only the proper adjustment and cleaning of the equipment can prevent the by-products of combustion.

It is generally true that the proper ratio of primary air to secondary air is the most important factor in maintaining the proper combustion of a fuel.

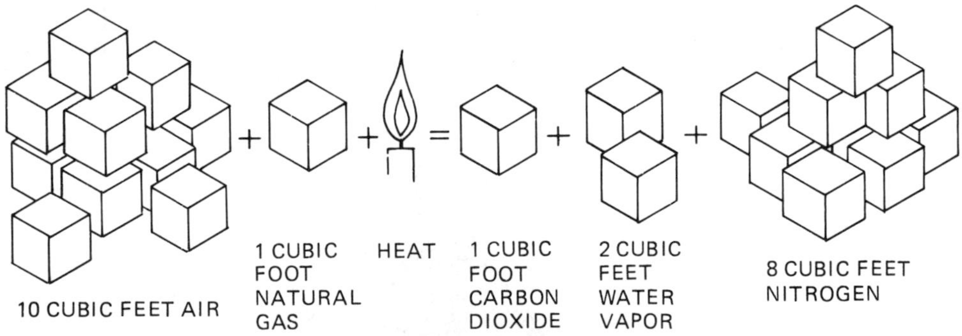

10 CUBIC FEET AIR    1 CUBIC FOOT NATURAL GAS    HEAT    1 CUBIC FOOT CARBON DIOXIDE    2 CUBIC FEET WATER VAPOR    8 CUBIC FEET NITROGEN

**FIGURE 5-2.**
Products of combustion

## Limits of Flammability

It is extremely important to remember the upper and lower limits of flammability when working with gaseous fuels. These limits are generally stated in percentages of fuel and air mixtures that will permit combustion. Table 5–1 indicates that if there is too much gas compared to air, the mixture will be too rich to burn. If there is too little gas compared to the air, the mixture will be too lean to burn.

When complete combustion is attained, all of the elements in the fuel will have combined with a sufficient amount of oxygen to be completely oxidized. When a fuel is completely burned the products of combustion are carbon dioxide ($CO_2$) and water vapor ($H_2O$), both of which are harmless.

## Combustion Rate

The combustion rate of the fuel depends upon three things: (1) the reaction of the fuel with the oxygen, (2) the rate that the oxygen is supplied to the fuel, and (3) the temperature in the combustion zone. In essence, the reaction rate is fixed by the type of fuel selected. However, the combustion rate can be increased by increasing either the mixing rate, the temperature, or both.

For complete combustion to occur, it is generally true that excess oxygen or excess air must be supplied beyond the theoretically required amount as a percentage of the air required to completely oxidize the fuel.

## Stoichiometric Combustion

Stoichiometric combustion is, theoretically, the ultimate in complete combustion. During this process the hydrocarbon fuel is combined with the exact amount of oxygen required to reduce all carbon, hydrogen, and sulfur

**TABLE 5–1**
Upper and Lower Limits of Fuel Gas Flammability

| Gas Type | Lower Limit | Upper Limit |
|---|---|---|
| Methane | 5.3 | 14 |
| Ethane | 3.2 | 12.5 |
| Natural | 3 | 14 |
| Propane | 2.4 | 9.5 |
| Butane | 1.9 | 8.5 |
| Manufactured | 4 | 29 |

in the fuel to $CO_2$, $H_2O$, and $SO_2$. Thus, the flue products theoretically contain only completely oxidized fuel elements or oxygen. The percentage of $CO_2$ is the maximum amount possible and is referred to as stoichiometric $CO_2$, ultimate $CO_2$, or maximum theoretical percentage of carbon dioxide.

Stoichiometric combustion is seldom realized because there is an infinite number of improper mixing rates of fuel and air, and there is an infinite number of reaction rates for each type of fuel under each circumstance. Because of these factors, the safety of the equipment and the desired economy demand that most equipment operate with at least some excess air. The excess air offers some guarantee that the fuel is not wasted. If the combustion equipment is properly designed, it will allow complete combustion even with varying grades of fuel and varying rates of supplying the air and fuel to the equipment. The best efficiency is obtained when the lowest amount of excess air possible is supplied to the equipment. Therefore, fuel burning equipment is designed to produce complete combustion and not stoichiometric combustion. The amount of excess air delievered to a given piece of equipment is governed by four factors:

1. Expected variations in fuel quality and the oxygen supply rates
2. Application of the equipment
3. The amount of manual operation required
4. The control requirements

## Incomplete Combustion

Incomplete combustion is dangerous, contributes to air pollution, is inefficient and must be avoided at all times. Incomplete combustion occurs when the fuel is not completely oxidized by the oxygen during the combustion process. As an example, a hydrocarbon element may not be completely reduced to carbon dioxide and water vapor. In this condition, the element can be changed into partially oxidized compounds, such as carbon monoxide, aldehydes, and keytones. Incomplete combustion is caused by the following conditions:

1. Improper mixing of the fuel and air
2. Insufficient amount of air supplied to the flame
3. Insufficient amount of time for the air and fuel to react to each other

4. The flame touching a cold surface

5. A low flame temperature

## Combustion Reactions

The amount of reaction between the oxygen and the combustible elements in the fuel occur in accordance with fixed chemical principles, which include the following:

1. A set of data that identifies the products of two or more reactants under a fixed set of conditions

2. The law of matter conservation (The mass of each element in the reaction products must equal the mass of that element in the reactants.)

3. The law of combining weights (Elements are combined in fixed weight relationships to form chemical compounds.)

The oxygen used in the combustion process is normally taken from the air introduced into the combustion zone, which is a physical mixture of nitrogen, oxygen, small amounts of water vapor, carbon dioxide, and inert gases, such as nitrogen and argon.

For practical combustion calculations, dry air is considered to consist of 20.95% oxygen and 79.05% inert gases by volume, or 23.15% oxygen and 76.85% inert gases by weight. In most calculation procedures, nitrogen is assumed to pass through the combustion process unchanged. Table 5-2 lists the oxygen and air requirements for stoichiometric combustion of some pure combustible materials found in common fuels.

## Heating Values

During the combustion process, heat energy is released. The amount of heat released by the complete combustion of a given fuel is constant. This amount of released heat is referred to as the heating value, the heat of combustion, or the calorific value of that particular fuel.

The heating value of a given type of fuel is determined by measuring the amount of heat produced during the complete combustion of a given quantity of that fuel in a calorimeter. Another method sometimes used is to

# TABLE 5-2
Combustion Reactions of Common Fuel Constituents

| Constituent | Molecular Symbol | Combustion Reactions | Stoichiometric Oxygen and Air Requirements | | | |
|---|---|---|---|---|---|---|
| | | | lb/lb Fuel[a] | | ft³/ft³ Fuel | |
| | | | $O_2$ | Air | $O_2$ | Air |
| Carbon (to CO) | C | $C + 0.5\alpha_2 \rightarrow CO$ | 1.33 | 5.75 | — | — |
| Carbon (to CO$_2$) | C | $C + O_2 \rightarrow CO_2$ | 2.66 | 11.51 | — | — |
| Carbon Monoxide | CO | $CO + 0.5\alpha_2 \rightarrow CO_2$ | 0.57 | 2.47 | 0.50 | 2.39 |
| Hydrogen | H$_2$ | $H_2 + 0.5\alpha_2 \rightarrow H_2O$ | 7.94 | 34.28 | 0.50 | 2.39 |
| Methane | CH$_4$ | $CH_4 + 2O_2 \rightarrow CO_2 + 2H_2O$ | 3.99 | 17.24 | 2.00 | 9.57 |
| Ethane | C$_2$H$_6$ | $C_2H_6 + 3.5O_2 \rightarrow 2CO_2 + 3H_2O$ | 3.72 | 16.09 | 3.50 | 16.75 |
| Propane | C$_3$H$_8$ | $C_3H_8 + 5O_2 = 3CO_2 + 4H_2O$ | 3.63 | 15.68 | 5.00 | 23.95 |
| Butane | C$_4$H$_{10}$ | $C_4H_{10} + 6.5O_2 = 4CO_2 + 5H_2O$ | 3.58 | 15.47 | 6.50 | 31.14 |
| — | C$_n$H$_{2n+2}$ | $C_nH_{2n+2} + (1.5n+0.5)O_2 \rightarrow nCO_2 + (n+1)H_2O$ | — | — | $1.5n+0.5$ | $7.18n+2.39$ |
| Ethylene | C$_2$H$_4$ | $C_2H_4 + 3O_2 \rightarrow 2CO_2 + 2H_2O$ | 3.42 | 14.78 | 3.00 | 14.38 |
| Propylene | C$_3$H$_6$ | $C_3H_6 + 4.5O_2 \rightarrow 3CO_2 + 3H_2O$ | 3.42 | 14.78 | 4.50 | 21.53 |
| — | C$_n$H$_{2n}$ | $C_nH_{2n} + 1.5nO_2 \rightarrow nCO_2 + nH_2O$ | 3.42 | 14.78 | $1.50n$ | $7.18n$ |
| Acetylene | C$_2$H$_2$ | $C_2H_2 + 2.5O_2 \rightarrow 2CO_2 + H_2O$ | 3.07 | 13.27 | 2.50 | 11.96 |
| — | C$_n$H$_{2m}$ | $C_nH_{2m} + (n+0.5m)O_2 \rightarrow nCO_2 + mH_2O$ | — | — | $n+0.5m$ | $4.78n+2.39m$ |
| Sulfur (to SO$_2$) | S | $S + O_2 \rightarrow SO_2$ | 1.00 | 4.31 | — | — |
| Sulfur (to SO$_3$) | S | $S + 1.5O_2 \rightarrow SO_3$ | 1.50 | 6.47 | — | — |
| Hydrogen Sulfide | H$_2$S | $H_2S + 1.5O_2 \rightarrow SO_2 + H_2O$ | 1.41 | 6.08 | 1.50 | 7.18 |

[a]Atomic masses: H = 1.008; C = 12.01; O = 16.00; S = 32.06.
(Courtesy of ASHRAE)

take a chemical analysis of the elements that make up the fuel and make an estimation of the heating value.

**Higher Heating Value or Gross Heating Value.** This property is determined when the water vapor in the products of combustion is condensed. In this method the latent heat of vaporization of water is included in the heating value of the fuel. Conversely, the lower heating value or net heating value is determined when the latent heat of vaporization is not included. When the heating value of a fuel is specified without designating the higher or lower, it generally means the higher heating value.

Fuel heating values are generally expressed in Btu/ft³ for gaseous fuels, Btu/gal for liquid fuels, and Btu/lb for solid fuels. The heating values are always given in relation to a specific reference temperature, usually 60, 68, or 77°F depending upon the practice in that particular industry. (See Table 5-3.)

When incomplete combustion occurs, the fuel is not completely oxidized, and the amount of heat released will be somewhat less than the recognized heating value for that fuel. Therefore, the amount of heat produced per unit of fuel consumed decreases, indicating a lower combustion efficiency.

**TABLE 5-3**
Heating Values of Substances Occurring in Common Fuels

| Substance | Molecular Symbol | Higher Heating Values[a] Btu/lb | Lower Heating Values[a] Btu/lb |
|---|---|---|---|
| Carbon (to CO) | C | 3950 | 3950 |
| Carbon (to $CO_2$) | C | 14,093 | 14,093 |
| Carbon Monoxide | CO | 4347 | 4347 |
| Hydrogen | $H_2$ | 61,095 | 51,023 |
| Methane | $CH_4$ | 23,875 | 21,495 |
| Ethane | $C_2H_6$ | 22,323 | 20,418 |
| Propane | $C_3H_8$ | 21,669 | 19,937 |
| Butane | $C_4H_{10}$ | 21,321 | 19,678 |
| Ethylene | $C_2H_4$ | 21,636 | 20,275 |
| Propylene | $C_3H_6$ | 21,048 | 19,687 |
| Acetylene | $C_2H_2$ | 21,508 | 20,769 |
| Sulfur (to $SO_2$) | S | 3980 | 3980 |
| Sulfur (to $SO_3$) | S | 5940 | 5940 |
| Hydrogen Sulfide | $H_2S$ | 7097 | 6537 |

[a]All values corrected to 60°F, 30 in. Hg, dry. For gases saturated with water vapor at 60°F, deduct 1.74% of the value.
(Courtesy of ASHRAE)

Not all of the heat released during the combustion process can be used. The greatest amount of heat lost is in the flue gases. Any increase in the temperature of the flue gases above the temperature of the incoming air and fuel represents a heat loss. Of course, some temperature rise is necessary in atmospheric draft combustion for operation of the venting system. Other losses include radiation and convection of heat transfer from the outer walls of the combustion equipment to the ambient air.

# FUEL CLASSIFICATION

Hydrocarbon fuels, generally, are classified according to their physical state (solid, liquid, gaseous). The type of combustion equipment for each fuel will be quite different. The gaseous types of fuels can be burned in premix or diffusion burners that take advantage of their gaseous state. The equipment for liquid fuels must include some means for atomizing or vaporizing the fuel into small droplets or a vapor for proper burning and for mixing the fuel and air. The equipment used for burning solid fuels must do the following:

1. Heat the fuel to vaporize a sufficient amount of volatiles to initiate and sustain combustion.
2. Provide enough time for the combustion process to be completed.
3. Have enough space to store the ashes from the fuel.

The type of fuel chosen will be based on one or more of the following:

1. Fuel factors
   a.  Availability, including dependability of supply
   b.  Convenience of use and storage
   c.  Economy
   d.  Cleanliness
2. Combustion equipment factors
   a.  Operating requirements
   b.  Cost
   c.  Service requirements
   d.  Ease of control

## Gaseous Fuels

In the past, various fuels have been used as sources of heat for boilers; however, today mostly natural and liquefied petroleum gases are used. The types and properties of the six most commonly used gaseous fuels are discussed below.

**Natural Gas.** Natural gas is an almost colorless and odorless gas that accumulates in the upper parts of oil and gas wells. Raw natural gas is a mixture of methane (55 to 98%), higher hydrocarbons (primarily ethane), and noncombustible gases. Some of the elements, principally water vapor, hydrogen sulfide, helium, liquefied petroleum gases, and gasoline are removed prior to distribution to the user.

The typical elements of the natural gas that is distributed to the user are: methane, $CH_4$ (70 to 96%); ethane, $C_2H_6$ (1 to 14%); propane, $C_3H_8$ (0 to 4%); butane $C_4H_{10}$ (0 to 2%); pentane, $C_5H_{12}$ (0 to 0.5%); hexane, $C_6H_{14}$ (0 to 2%); carbon dioxide, $CO_2$ (0 to 2%); oxygen, $O_2$ (0 to 1.2%), and nitrogen, $N_2$ (0.4 to 17%). Natural gases are sometimes divided into three types: (1) high inert, (2) high methane, and (3) high Btu. (See Table 5–4.)

The composition of natural gas depends upon the geographical area from which it is taken. Therefore, the composition of the gas will vary from location to location but a fairly constant heating value is usually maintained for control and safety purposes. The local gas company is usually the best source for determining the composition for any particular area.

The heating values of natural gases vary from 900 to 1200 Btu/ft³, with the most usual ranging from 1000 to 1050 Btu/ft³. Unknown heating values can be calculated for particular gases by using the composition data and Table 5–3.

**TABLE 5–4**
Group Classifications of Natural Gases

| Group | Nitrogen, % | Specific Gravity | Methane, % | Btu/ft³ Dry |
|---|---|---|---|---|
| I High inert type | 6.3 to 16.20 | 0.660 to 0.708 | 71.9 to 83.2 | 958 to 1051 |
| II High methane type | 0.1 to 2.39 | 0.590 to 0.614 | 87.6 to 95.7 | 1008 to 1071 |
| III High Btu type | 1.2 to 7.5 | 0.620 to 0.719 | 85.0 to 90.1 | 1071 to 1124 |

(Courtesy of ASHRAE)

Odorants, such as mercaptans, are added to natural gas to give it a characteristic odor for safety purposes.

### Liquefied Petroleum Gases.

Liquefied petroleum gases are made up of primarily propane and butane. It is normally obtained as a by-product of oil refinery operations or by stripping natural gas. Both propane and butane are gaseous under normal atmospheric conditions. However, they can be liquefied by the application of moderate pressures at normal temperatures.

Three liquefied petroleum gases are commercially available as heating fuels: (1) butane, (2) propane, and (3) a mixture of the two.

### Commercial Propane.

The commercial grade of propane is made up of propane along with about 5 to 10% propylene. It is rated with a heating value of about 21,560 Btu/lb or about 2500 Btu/ft$^3$ of gas. It has a boiling point of about $-40°F$ at atmospheric pressure. Propane, because of its low boiling temperature, is especially useful during the winter months where a low outdoor temperature is probable. It is available in bottles, tanks, bulk, cylinders, and railroad car loads.

### Commercial Butane.

Commercial butane is basically made up of butane which contains up to 5% butalene. It is rated with a heating value of about 21,180 Btu/lb or 3200 Btu/ft$^3$ of the gas. Commercial butane at atmospheric pressure has a boiling temperature of 32°F. Because of this relatively high boiling temperature, butane is not a very good fuel in the colder climates because it will not boil off and produce a vapor to be burned in the equipment. In almost all areas a mixture of butane and propane is used to lower the boiling temperature of the mixture so that it can be used in cold weather. Butane is available in bottles, bulk, and tank car loads.

### Commercial Propane-Butane Mixtures.

Commercial propane-butane mixtures are available during the winter months with varying ratios of the mixture to meet the local existing temperatures. The heating value and all the other properties fall between the values for the pure form of each type, depending upon the percentage of each mixture.

Also available are propane-air and butane-air mixtures which are used in place of natural gas in some of the smaller communities, and during peak load periods by natural gas companies. Each of the various fuel-air ratios has different heating values and specific gravities. (See Table 5–5.)

### Manufactured Gases.

Manufactured gases are gases which are produced from coke, coal, oil, liquefied petroleum gases or natural gas. These types

**TABLE 5–5**
Propane-Air and Butane-Air Gas Mixtures

| | Propane-Air[a] | | | Butane-Air[b] | | |
|---|---|---|---|---|---|---|
| Btu/ft³ | %Gas | %Air | Sp Gr | %Gas | %Air | Sp Gr |
| 500 | 19.8 | 80.2 | 1.103 | 15.3 | 84.7 | 1.155 |
| 600 | 23.8 | 76.2 | 1.124 | 18.4 | 81.6 | 1.186 |
| 700 | 27.8 | 72.2 | 1.144 | 21.5 | 78.5 | 1.216 |
| 800 | 31.7 | 68.3 | 1.165 | 24.5 | 75.5 | 1.248 |
| 900 | 35.7 | 64.3 | 1.185 | 27.6 | 72.4 | 1.278 |
| 1000 | 39.7 | 60.3 | 1.206 | 30.7 | 69.3 | 1.310 |
| 1100 | 43.6 | 56.4 | 1.227 | 33.7 | 66.3 | 1.341 |
| 1200 | 47.5 | 52.5 | 1.248 | 36.8 | 63.2 | 1.372 |
| 1300 | 51.5 | 48.5 | 1.268 | 39.8 | 60.2 | 1.402 |
| 1400 | 55.5 | 44.5 | 1.288 | 42.9 | 57.1 | 1.433 |
| 1500 | 59.4 | 40.6 | 1.309 | 46.0 | 54.0 | 1.464 |
| 1600 | 63.4 | 36.6 | 1.330 | 49.0 | 51.0 | 1.495 |
| 1700 | 67.4 | 32.6 | 1.350 | 52.1 | 47.9 | 1.526 |
| 1800 | 71.3 | 28.7 | 1.371 | 55.2 | 44.8 | 1.557 |

[a]Values used for calculation: 2522 Btu/ft³; 1.52 specific gravity.
[b]Values used for calculation; 3261 Btu/ft³; 2.01 specific gravity.
(Courtesy of ASHRAE)

of fuels are used primarily for industrial in-plant operations or as specialty fuels.

## Liquid Fuels

Liquid fuels, for the most part, are mixtures of hydrocarbons which are obtained by refining crude oil. They almost always contain small quantities of sulfur, nitrogen, oxygen, vanadium, other trace metals, and impurities such as water and sediment. During the refining process other lightweight petroleum products and fuels such as gasoline, butane, propane, kerosene, jet fuels, diesel fuels, and lightweight heating oils are produced. There are heavy hydrocarbons produced during the refining process such as lubricating oils, waxes, petroleum coke and asphalt.

The crude oil obtained from different drilling fields varies in hydrocarbon molecular structure. Crude oil has several different types of bases, such as: (1) paraffin base, (2) naphthalene or asphaltic base, (3) aromatic base, or (4) mixed or intermediate base. Except in the case of heavy fuel oils, the type of crude has very little effect on the refined products or the combustion characteristics of the fuel.

**Fuel Oil Types.**   Heating fuel oils have two classifications: (1) distillate, or lighter fuel oils, and (2) residual, heavier fuel oils. The American Society for Testing and Materials (ASTM) has set forth specifications for fuel oil properties which include five grades of fuel oil (grades 1 through 6 omitting number 3), as follows:

*Grade Number 1.*   This is a light distillate which is intended for use in vaporizing-type burners. High volatility is required to continue the evaporation process of the fuel oil with a minimum amount of residue.

*Grade Number 2.*   This is a heavier grade of fuel oil than distillate number 1. It is to be used with pressure-atomizing (gun) burners that spray the oil into the combustion chamber. The atomized oil vapor mixes with the air and burns. This is the type that is used in most domestic fuel oil burners and many medium-capacity, commercial-industrial type burners.

*Grade Number 4.*   This is an intermediate fuel oil and is considered to be either a light residual or a heavy distillate-type fuel oil. Its main use is in atomizing-type burners designed for fuel oils of a heavier viscosity than most domestic oil burners can handle. Its permissible viscosity range allows it to be pumped and atomized at relatively low storage temperatures.

*Grade Number 5 (Light).*   This is a light residual type of fuel oil which has an intermediate viscosity. It is used in burners that are designed for use with a fuel that is more viscous than the number 4 type without preheating. Preheating of this fuel may be required for burning in some types of equipment and for handling in colder climates.

*Grade Number 5 (Heavy).*   This is a heavy residual type of fuel oil which is more viscous than the number 5 (Light). It is however, intended for very similar types of service. Preheating is usually needed in colder climates for both burning and handling purposes.

*Grade Number 6.*   This grade of fuel oil is sometimes referred to as bunker C oil. It has a high viscosity and is used mostly in commercial and industrial applications. Preheating is required in the storage tank to permit pumping of the oil. Additional preheating is also needed at the burner to properly atomize it for burning.

In some areas low sulphur residual fuel oils are produced to permit users to meet the emission regulations for sulfur dioxide. These types of fuel oils are produced by the following methods:

1. By refinery processes which removes the sulfur from the fuel oil (a process known as hydrodesulfurization)

2. By blending high sulfur residual oils with low sulfur distillate oils
3. By a combination of all of these methods

Because of these refining procedures, these types of fuel oils have different characteristics than the other types of residual fuel oils. For example, the viscosity-temperature relationship may be such that low sulfur oils have viscosities of number 6 oil when cold and those of the number 4 when heated. Therefore, some alteration of the normal guidelines for fuel oil burning and handling is usually possible.

## Solid Fuels

Some of the more popular types of solid fuels are coal, coke, wood, and waste products from industrial and agricultural operations. Coal is probably the most popular of all these solid fuels.

The classification of coal is relatively difficult because of its rather complex composition. Coal consists of carbon, hydrogen, oxygen, nitrogen, sulfur, and mineral residue called ash. A chemical analysis provides an indication of its quality. However, this analysis does not completely indicate its burning characteristics. The user is mainly interested in the amount of available heat per pound of the coal, the amount of ash produced, the handling and storing requirements, and the burning characteristics of the type of coal under consideration.

**Coal Types.**    There are some accepted definitions that are used when classifying coal types. (See Table 5–6.) The classification of coal is rather arbitrary however, because there are no distinct dividing lines between the different types of coals in use.

*Anthracite.*    Anthracite coal is a hard, clean, dense type which creates very little dust when it is handled. It is rather hard to ignite, but it does burn easily after it is lighted. It burns uniformly with a short, smokeless flame and its ash is noncaking.

*Semianthracite.*    Semianthracite is similar to anthracite coal but with the following exceptions: it is not as hard, it is easier to start burning, and it has a higher volatile content.

*Semibituminous Coal.*    Semibituminous is a soft and crumbly type of coal. Therefore, a fine dust is created when handling it. It is slow to start burning and it burns with a medium length flame. As the amount of volatile

**TABLE 5-6**

Classification of Coals by Rank[a]

Legend: F.C. = Fixed Carbon.   V.M. = Volatile Matter.

| Class | Group | Limits of Fixed Carbon or Btu Mineral-Matter-Free Basis | Requisite Physical Properties |
|---|---|---|---|
| I. Anthracite | 1. Meta-anthracite | Dry F.C., 98% or more (Dry V.M., 2% or less) | Nonagglomerating |
| | 2. Anthracite | Dry F.C., 92% or more, and less than 98% (Dry V.M., 8% or less, and more than 2%) | |
| | 3. Semianthracite | Dry F.C., 86% or more, and less than 92% (Dry V.M., 14% or less, and more than 8%) | |
| II. Bituminous[d] | 1. Low-volatile bituminous coal | Dry F.C., 78% or more, and less than 86% (Dry V.M., 22% or less, and more than 14%) | Either agglomerating[b] or nonweathering[f] |
| | 2. Medium-volatile bituminous coal | Dry F.C., 69% or more, and less than 78% (Dry V.M., 31% or less, and more than 22%) | |
| | 3. High-volatile A bituminous coal | Dry F.C., less than 69% (Dry V.M., more than 31%), and moist[c], about 14,000 Btu[e] or more | |
| | 4. High-volatile B bituminous coal | Moist[c], about 13,000 Btu or more, and less than 14,000[e] | |
| | 5. High-volatile C bituminous coal | Moist, about 11,000 Btu or more, and less than 13,000[e] | |

174

| III. Subbituminous | 1. Subbituminous A coal | Moist, about 11,000 Btu or more, and less than 13,000[e] | Both weathering and non-agglomerating[b] |
|---|---|---|---|
| | 2. Subbituminous B coal | Moist, about 9500 Btu or more, and less than 11,000[e] | |
| | 3. Subbituminous C coal | Moist, about 8300 Btu or more, and less than 9500[e] | |
| IV. Lignitic | 1. Lignite | Moist (Btu) less than 8300 | Consolidated |
| | 2. Brown coal | Moist (Btu) less than 8300 | Unconsolidated |

Adapted from *ASTM Standards*, 1937, *Supplement*, p. 145, *American Society for Testing and Materials.*

[a]This classification does not include a few coals of unusual physical and chemical properties which come within the limits of fixed carbon or Btu of high-volatile bituminous and subbituminous ranks. All these coals either contain less than 48% dry, mineral-matter-free fixed carbon, or have more than about 15,500 moist, mineral-matter-free Btu.

[b]If agglomerating, classify in low-volatile group of the bituminous class.

[c]Moist (Btu) refers to coal containing its natural bed moisture but not including visible water on the coal surface.

[d]There may be noncaking varieties in each group of the bituminous class.

[e]Coals having 69% or more fixed carbon on the dry, mineral-matter-free basis shall be classified according to fixed carbon, regardless of Btu.

[f]There are three varieties of coal in the high-volatile C bituminous coal group: Variety 1, agglomerating and nonweathering; Variety 2, agglomerating and weathering; Variety 3, nonagglomerating and nonweathering.

(Courtesy of ASHRAE)

matter is increased, its caking properties are also increased, but the coke formed by its burning is weak. It has only half the volatile matter of bituminous coal. During the burning process less smoke is produced. It is sometimes called smokeless coal.

*Bituminous Coal.* Bituminous coal is made up of several different types of coal having different compositions, properties, and burning characteristics. Bituminous coal from the east is a higher grade than that from the west. The different types of bituminous coal have different caking qualities, from the melting, plastic type to the types that are classified as noncaking or free-burning. Most coals of the bituminous type are considered to be noncrumbling. Bituminous coals are generally easy to ignite and they burn easily. Generally they have a long flame length which varies with the different types. Bituminous coals have a tendency to smoke and produce soot when not properly fired, especially at low burning rates.

*Subbituminous Coal.* Subbituminous coals are high in moisture content and have a tendency to crumble as they are dried or are exposed to the weather. This is especially true of those mined in the west. When they are stored in piles, spontaneous combustion is possible. Subbituminous coals are easy and quick to ignite. They are free-burning, have a medium flame length, and are noncaking. Any lumps will break into small pieces when poked. There is very little smoke or soot formed when these coals are burned.

*Lignite Coal.* Lignite coal is considered to be very woody in nature. It has a very high moisture content when first mined. Lignite is clean to handle, but it has a low heating value. During the drying process, it has a greater tendency to disintegrate than subbituminous types of coal. Also, spontaneous combustion is more likely when lignite coal is stored than when subbituminous coal is stored. Lignite is noncaking, it is slow to ignite when first mined. The char is easily burned after the moisture and volatile matter have been driven away. The remaining lumps have a tendency to break up in the fuel bed and any char that may fall into the ashpit will continue to burn. Lignite forms very little smoke or soot when burning.

## Combustion Considerations

There are practical combustion considerations that should be given to any piece of fuel-burning equipment. Several of these are described below.

**Air Pollution.**    During the combustion process, certain pollutants are released to the atmosphere. These pollutants are the largest single source of air pollution. They can be grouped into four categories.

1. Products of incomplete combustion
   a. Combustible aerosols (solid and liquid) including smoke, soot, and organics; excluding ash
   b. Carbon monoxide (CO)
   c. Gaseous hydrocarbons (HC)
2. Oxides of nitrogen (generally grouped and referred to as $NO_x$)
   a. Nitric oxide (NO)
   b. Nitrogen oxide ($NO_2$)
3. Emissions resulting from fuel contaminants
   a. Sulfur oxides, primarily sulfur dioxide ($SO_2$), and small quantities of sulfur trioxide ($SO_3$)
   b. Ash
   c. Trace metals
4. Emissions resulting from additives
   a. Combustion-controlling additives
   b. Other additives

The release of the products of incomplete combustion and nitrogen oxides are directly related to the combustion process and can be properly controlled by making proper adjustments to the equipment. The release of fuel contaminants is related to the type of fuel selected and is only slightly affected by the combustion process. The additives that are added to the fuel to obtain certain characteristics also create certain emissions. When these additives are to be used, their impact on the pollution problem should be considered and evaluated in relation to the amount of good that they can contribute.

The release of the products of incomplete combustion can be reduced by properly adjusting the equipment to ensure that the most complete combustion possible is attained. Some of these adjustments are (1) providing sufficient excess air, (2) improving the air- and fuel-mixing process, and (3) increasing the amount of time that the combustible mixture remains inside the furnace.

Nitrogen oxides are produced during the combustion process by one of two ways: (1) thermal fixation (reaction of the nitrogen and oxygen at high combustion temperatures), or (2) from nitrogen in the fuel (oxidation of organic nitrogen in the fuel molecules). It should be noted that the techniques that are used to ensure complete combustion also tend to promote increased $NO_x$ formation. These techniques make use of low excess-air firing and firing at higher flame temperatures. The only technique that has reduced the $NO_x$ is two-stage firing. In two-stage firing, the lean, air-deficient, primary air combustion zone tends to prevent the formation of $NO_x$ during the low firing rate. This is the time that $NO_x$ forms the easiest from the nitrogen in the fuel. Also, the amount of peak temperatures are reduced which in turn reduces thermal fixation.

$NO_x$ control procedures are widely used on large boilers, but are not used on the smaller types of combustion equipment.

**Condensation and Corrosion.** Sulfuric acid, which is produced when the sulfur trioxide and the water vapor contained in flue gases combine, causes corrosion of the venting system. It occurs during periods of operation with low temperature flue gases, which permits condensation of the flue gases. It is usually recommended that the flue gas temperature be kept above the acid dew point temperature from the point of combustion through the venting system.

Corrosion is also sometimes caused by sodium, potassium, chlorine, and vanadium. All of these are found in residual fuel oils and in the different types of coal.

**Soot.** Soot that has collected on the heat exchanger surfaces acts as an insulator and reduces the amount of heat transferred from the flame to the water or steam. It also clogs the flue passages and venting system, thereby reducing the draft through the equipment and interfering with the combustion process. In most instances, the proper adjustment of the burner can keep soot from forming. Once it has formed, it is usually safer and better to completely clean the equipment manually. The aerosol sprays that are available will only add to the corrosion problem.

# REVIEW QUESTIONS

1. Define combustion.
2. Name the three basic requirements for complete combustion.
3. How can the by-products of combustion be prevented?

4. What could happen to a gaseous flame if there was too little air supplied to the gas?

5. When does incomplete combustion occur?

6. What happens to the nitrogen in the air during the combustion process?

7. What does the term *heat of combustion* of a fuel refer to?

8. What property of a fuel is determined when the water vapor in the products of combustion is condensed?

9. When is the lower heating value or the net heating value of a fuel determined?

10. How does incomplete combustion effect efficiency?

11. What are the two most popular types of gaseous fuels used in boilers?

12. What do the terms *high inert, high methane,* and *high Btu* describe?

13. What are manufactured gases used for?

14. What are liquid fuels?

15. What are the two classifications of heating fuel oils?

16. What grade of fuel oil is also called bunker C oil?

17. Waste products from industrial and agricultural operations are what type of fuel?

18. Name the four categories of air pollutants produced by boiler combustion.

19. In what two types of fuel are sodium, potassium, chlorine, and vanadium found?

20. Will a collection of soot on the heat exchanger surface affect the heat transfer process?

# CHAPTER 6

# COMBUSTION CONTROLS

## Learning Objectives

The objectives of this chapter are to:

- Introduce the reader to the basic combustion controls used on boilers
- Introduce the high-low fire operation of a boiler
- Describe the operation of flame failure and operating controls
- Show the reader the trial-for-main-burner-ignition operation.

Combustion control is the process of regulating the mixed flow of air and fuel to a boiler furnace to supply the needed amount of steam or hot water. Combustion controls help the boiler technician maintain the proper burning rate of the fuel. The four things that are regulated by the combustion control are (1) air supply, (2) fuel supply, (3) the ratio of fuel to air, and (4) flame safety. Each of these functions is described below.

## AIR SUPPLY

As we learned earlier, two different types of combustion air are needed in order for proper combustion to occur. The primary air is mixed with the fuel before ignition occurs, and the secondary air is forced into the boiler furnace to help complete the combustion process and control the efficiency of the flame.

## FUEL SUPPLY

The fuel supply is regulated by the demand for steam or hot water. When the demand for heating is great the burner operates in the high-fire mode. When the demand for heating is small the burner operates in the low-fire mode.

## THE RATIO OF FUEL TO AIR

When the demand for heat increases the burner will go into the high-fire mode of operation, and when the demand is small the burner will go into the low-fire mode. In each of these modes of operation a different amount of combustion air is required. The combustion air dampers are actuated according to the fuel demand. The combustion air control causes the air damper to open and allow more primary air and secondary air into the combustion zone when the demand is high. When there is a drop in the demand, the combustion air control causes the air damper to partially close, reducing the amount of combustion air to the combustion zone.

# SAFETY

The safe operation of a boiler is just as important as efficient operation. The safety controls insure that the boiler does not operate when an unsafe condition exists. Most boilers use the electronic type of safety control because they react much faster than thermocouple-type controls to an unsafe condition.

# AUTOMATIC CONTROLS

Automatic controls are a big asset because they reduce the amount of manual control required for the boiler. They also aid in the combustion process and in operation of the auxiliary equipment.

## Air and Fuel Control

When the operating control demands that the operation of the boiler be changed, a signal is received by a modulating type of motor. The change indicated can be to start the boiler or turn it off; or to change it from high-fire to low-fire operation, or from low-fire to high-fire operation. The modutrol motor starts to turn which, in turn, opens or closes the fuel valve and the air shutters in order to adjust the firing rate to suit the demands on the boiler. (See Figure 6–1.)

The modulating motor consists of the motor windings, a balancing relay, and a balancing potentiometer. The loading is transmitted to the winding through an oil-immersed gear train from the crank arm. The crankshaft is the double-ended type, and the crank arm may be mounted on either end of the motor. The motor works in conjunction with the potentiometer coil in the modulating operating control. An electrical imbalance is created by either a change in the temperature of the water or a change in the steam pressure, which causes the operating control to make a demand on the boiler. This causes the motor to rotate in an attempt to rebalance the circuit. The crank arm, through linkage, positions the burner air louvers and the fuel regulating valve, maintaining a balanced flow of air and fuel throughout the burner firing range.

**FIGURE 6-1.**
Modulating motor (Courtesy of Johnson Controls, Inc., Control Products Div.)

## Flame Failure and Operating Controls

Frequently, on full-automatic boilers, an electronic type of device is provided for the control of flame failure. The device will provide automatic start and operation of the main burner equipment. The controls are so designed that they close all fuel valves, shut down the burner equipment within four seconds after a flame failure, and actuate an alarm. The controls also create a safety shutdown within four seconds after de-energization of the ignition equipment if the main burner flame is not properly established, or fails during the normal starting sequence. The controls must create a safety shutdown if the pilot flame is not established and confirmed within seven seconds after lighting. A safety shutdown requires manual reset before operation can be resumed and prevents recycling of the burner equipment.

In all initial starts and subsequent restarts during on-and-off cycling, the automatic operation controls are designed to provide the services described below.

**Low-Fire Start.** Almost all boilers are operated with a low-fire start, especially when they are cold. They will operate in this mode until the boiler has reached a temperature high enough to prevent thermal shock, which may result from bringing it up to operating temperature too fast.

**Precombustion and Postcombustion Scavenging.** Precombustion and postcombustion scavenging is used on most boilers over the size of 2,000 lb/hr of steam. Boilers smaller than this do not require the postcombustion scavenging operation. During the precombustion and the postcombustion scavenging periods the combustion air blower should be on, and the ignition, pilot, and main fuel valve should be off.

**Time Delay.** This control provides a time delay of at least five seconds between the ignition of the pilot and the opening of the main fuel valve, except in gas-fired boilers and light oil-fired boilers with a pilot. In these, the pilot flame must be established and confirmed before the main fuel valve can be opened.

**Trial-for-Main-Burner-Ignition.** The trial-for-ignition is the time that the pilot flame and/or ignition is on after the main fuel valve opens. After the prescribed period of time has elapsed and the main burner flame has been properly established and confirmed, there will be a proper shift from the operating control to the combustion operating controls.

In the case of supply voltage failure, the operating control will recycle when the power is restored. The electronic device should readily detect a flame under all firing conditions of both the pilot and the main flame, but it should not be actuated by a hot refractory or any other hot body. The controls are designed so that a failure of any vacuum tube or other component will cause a safety shutdown, or result in a safety interlock to prevent recycling of the control.

A typical and modern combustion safety device is the electronic type using a photocell (or electric eye) to detect the presence of a flame. (See Figure 6–2.) This is a fast-acting device and is referred to by such names as flame eye, scanner, and electric eye. The photocell reacts to the light of the flame, thereby allowing electric current to flow in the electronic circuit and hold the flame relay in a closed position. A failure of the flame interrupts

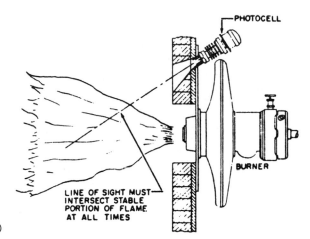

**FIGURE 6-2.**
Photocell (electric eye)

the electronic circuit, causing the burner to shut down on safety. The photocell is provided with a time delay of 4 to 8 seconds in order to prevent nuisance shutdowns due to flame fluctuations, especially at low-fire positions.

**Steam- and Air-Flow Meter.**    A combustion air- and steam-flow meter is used as a guide in controlling the relationship between the air required and the air actually supplied to burn the fuel. (See Figure 6–3.) The rate of steam generation is used as a measure of the air necessary to burn the required amount of fuel. The flow of gases through the boiler furnace is used as a measure of the air supplied.

Essential parts of the meter are two air-flow bells supported from knife edges on a beam, which is supported by other knife edges, and a mercury displacer assembly supported by a knife edge on the beam. The bottoms of the bells are sealed with oil and the spaces under the bells are connected to two points of the boiler setting.

**Linkage (Air and Fuel Adjustments).**    The purpose of the linkage is to regulate the amount of combustion air in correct proportion to the amount of fuel required to obtain the burner's desired firing rate at maximum efficiency. To obtain the desired results, the linkages can be manually adjusted at various points between the air damper and the modulating fuel valves.

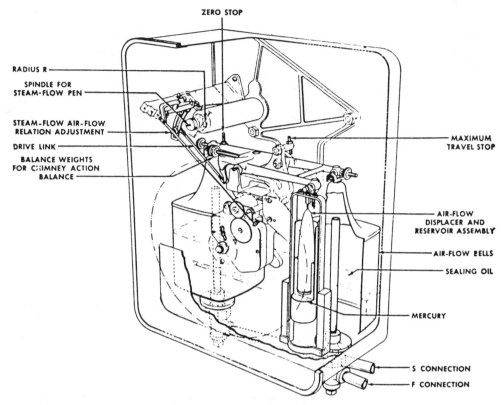

**FIGURE 6-3.**
Air-flow mechanism of a boiler meter

# PROGRAMMING CONTROLS

The purpose of the programming control is to oversee operation of the boiler burner. When the operating control demands heat, the programming control signals all of the fuel valves, fuel oil heaters, the oil burner and any other devices used on a particular installation. If the main burner flame is not proven in a given period of time the programming control will automatically shut down the boiler burner. It allows for pre-purge time so that all the combustible products are blown from the boiler furnace.

At given intervals, certain functions are monitored. The most commonly monitored functions are pre-purging, proving combustion air flow, initiating low-fire/high-fire operation, energizing and de-energizing the ignition transformer and pilot solenoid valve, proving the pilot flame, and monitoring the main burner flame.

When the demand has been satisfied, the programming control shuts down the boiler, and provides for a post-purge period. After the post-purge cycle the programming control is usually ready for the next operating cycle.

There are so many different makes and models of programming controls that it would be impossible to give the operating description of each of them in a manual of this type. For more specific operating instructions for the particular model in question, refer to the manufacturer's instructions packaged with the control.

# REVIEW QUESTIONS

1. What two types of air are used in the combustion process?
2. On a boiler, what controls the amount of combustion air to the boiler furnace?
3. Why are the electronic type of controls more popular than thermocouple-type controls?
4. What causes a modulating motor to turn either one way or the other?
5. When electronic controls are used on a boiler, within what time frame will the system shut down if the main burner flame is not properly established?
6. In what firing mode are almost all boilers started?
7. During the precombustion and postcombustion scavenging periods, is the pilot valve energized by the electronic control?
8. What is the normal time delay between the pilot ignition and the opening of the main fuel valve on most boilers?
9. What should the electronic control *not* be actuated by?
10. What does a photocell react to?

# CHAPTER 7

# DRAFT CONTROL

## Learning Objectives

The objectives of this chapter are to:

- Show the different types of draft used in boiler operation
- Acquaint the reader with the different components used in boiler venting
- Describe the different methods of controlling boiler draft flow
- Show reasons for measuring the draft through a boiler
- Introduce the reader to the different components of a boiler draft.

As we learned in Chapter 5, combustion requires that a minimum amount of air be supplied to the boiler furnace during the combustion process so that the maximum amount of heat can be received from the burning fuel. The amount of air required depends on the type of fuel being burned and whether the boiler is operating in the high-fire or the low-fire mode. The air going into the boiler furnace, along with the amount of gases that the burning fuel generates, must be discharged to the atmosphere to make room for the incoming fuel and air. This venting of the products of combustion is done by four different methods: (1) natural draft, (2) forced draft, (3) induced draft, and (4) combination forced draft and induced draft.

# NATURAL DRAFT

Natural draft occurs when there is a difference in pressure between two places. The air moves from the place with the higher pressure to the place with the least pressure. Natural draft burners are not generally used in the larger systems because there is not sufficient airflow through the furnace to support the combustion process. Also, proper control over the combustion air cannot be maintained to produce the proper combustion. A natural draft burner does not respond fast enough for modern day high-capacity boiler burners.

# FORCED DRAFT

A forced-draft system has an electric blower mounted so that it moves enough air to keep the furnace under a slight pressure. (See Figure 7–1.) The air is forced into the boiler furnace by the fan. The air then passes through the fuel bed for coal, or through the burner for oil or gas, where it is thoroughly mixed. Then the air passes into the combustion zone where it aids in completing the combustion process. As the air is forced into the furnace, the flame becomes hotter because the fuel is being burned more completely. The products of combustion then pass from the boiler into the breeching, out of the breeching, and into the chimney. The purpose of the chimney is to help produce the draft and to convey the products of combustion to the atmosphere, where they are harmless. The amount of air supplied to the boiler is determined by the demand for hot water or steam from the boiler. A good chimney is needed to carry the products of combustion when a

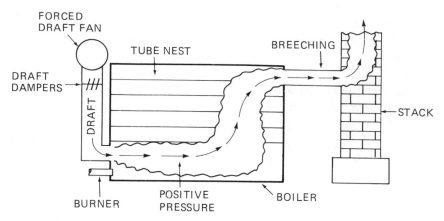

**FIGURE 7-1.**
Forced draft through a boiler

forced-draft system is used. Otherwise, the products of combustion may be forced into occupied areas.

## INDUCED DRAFT

An induced draft uses an electric blower located in the breeching of the boiler. (See Figure 7–2.) An induced draft in a boiler furnace causes the furnace to be under a slight negative pressure. The combustion air passes

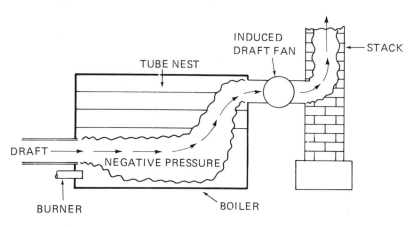

**FIGURE 7-2.**
Induced draft through a boiler

through the boiler furnace just as it does when a forced-draft system is used. However, the furnace operates under a slightly negative pressure because the air is drawn in rather than blown in. Therefore, the furnace must be airtight to prevent air being drawn into the system when it is not needed to aid in the combustion process. The products of combustion are blown by the blower into the chimney and then out into the atmosphere. When induced draft is used, the chimney does not need to be quite as airtight as when the forced-draft system is used.

## COMBINATION FORCED AND INDUCED DRAFT

In some installations both forced draft and induced draft are used to achieve the desired results. When this method is used, blowers are located both at the inlet of the furnace and in the breeching. (See Figure 7–3.) This type of system has most of the advantages of both the forced-draft and the induced-draft systems. However, combination draft systems are only used in special applications because of their cost.

## BREECHINGS

Breechings are used to connect the boiler to the chimney. They are generally made of sheet steel, with provisions for expansion and contraction. The

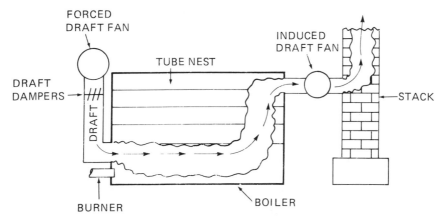

**FIGURE 7-3.**
Combination forced draft and induced draft through a boiler

breeching may be carried over the boilers, in back of the furnace, or even under the boiler room floor. The breechings should be kept as short as possible, and should be free from sharp bends and abrupt changes in area. The cross-sectional area should be approximately 20% greater than that of the chimney to keep draft loss to a minimum. A breeching with a circular cross section causes less draft than one with a rectangular or square cross section.

# CONTROLLING THE DRAFT

Automatic draft control is needed with forced-draft boilers only when they must vent their combustion products into a chimney that produces an excessive amount of natural draft or negative draft from an induced-draft unit. In a multistory building, for example, an existing chimney may produce an extremely high draft. When several boilers are served by a single chimney, the draft may be satisfactorily low when all of the boilers are operating, but it may become excessive when only one or two of the boilers are operating. In these and other cases where a variable negative draft may exist at the boiler outlet, sequence-type overfire draft controls or barometric dampers are the best choices.

The first choice is the sequence-type controls, which maintain the desired positive pressure inside the furnace regardless of any negative pressure that may exist at the inlet to the breeching. Barometric dampers help reduce negative draft but do not provide the degree of control required for the best burner performance.

## Barometric Draft Controls

When barometric draft controls are installed, they should have a minimum distance of one flue pipe diameter from the top of the boiler jacket to the centerline of the draft control. Ideally, they should be mounted in the end of a vent pipe tee. When mounted in this position they have better control over minor down drafts. (See Figure 7–4.) (NOTE: Barometric draft controls should always be installed in the same room as the boiler. Never install them in another room.)

**Draft Control Operation.**   The boiler vent draft is extremely susceptible to atmospheric conditions because the temperature, humidity, and wind velocity of the atmosphere changes the amount of natural draft through the fur-

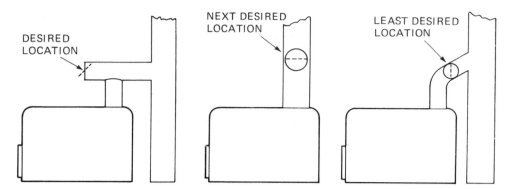

**FIGURE 7-4.**
Location of barometric draft control

nace of the boiler. This also changes the amount of draft which passes through the chimney. The purpose of the barometric draft control is to compensate for these changes in the vent draft.

**Barometric Draft Control Operation.**  In operation, the barometric draft control damper reduces the amount of heat lost up the chimney because of excessive drafts. This is accomplished by allowing some of the vent draft to be drawn into the venting system through the damper rather than all of it being drawn through the firebox of the boiler. This operation permits an exact amount of air to be supplied to the fuel for proper burning, which increases the efficiency of the boiler.

**Barometric Draft Control Adjustment.**  The draft control may be adjusted to allow the desired amount of draft through the boiler furnace. These adjustments should be made in accordance with the manufacturer's recommendations. Generally, they should be adjusted to provide a carbon dioxide ($CO_2$) reading between 8% and 9.5% with a maximum carbon monoxide (CO) of 0.025% and a draft reading of between $-0.02''$ water column and $-0.04''$ water column. This is determined by the actual $CO_2$ and the CO reading taken at the boiler flue.

## Measuring the Draft

The draft is measured in the breeching after the boiler has been in operation long enough for the complete system to be warmed up to the normal operating temperature. The instrument used to check the draft is either a U-tube or a draft gauge.

**Draft Gauges.**   A draft gauge is a form of pressure gauge. In boiler work, the term *draft* usually refers to the pressure difference producing the flow. Drafts are pressures which are below atmospheric pressure. They are measured in inches of water column. A draft gauge is essential to safe and economical boiler operation.

A simple type of draft gauge is the U-tube gauge. (See Figure 7–5.) The source of draft is connected to one leg of the U and the other end is left open to the atmosphere. The difference between the levels of liquid in the two legs is a measure of the draft. Water is generally used in this type of gauge. Take a close look at Figure 7–6, which shows a comparison of an inclined-draft gauge and a U-tube gauge.

If one leg of the U-tube is arranged on an incline, the distance moved by the liquid in the inclined portion is increased for a given draft change, which makes it possible to obtain more accurate readings.

Two or more draft gauges are required for economical boiler operation. They inform the operator of the relative amount of air being supplied to burn the fuel and the condition of the flue gas passages. Draft gauges are made as indicators, recorders, or both. The measuring element uses a column of liquid, a diaphragm, or a bellows. The liquids used are oil, water, or mercury.

There is an indicating type of gauge that operates on the same principle

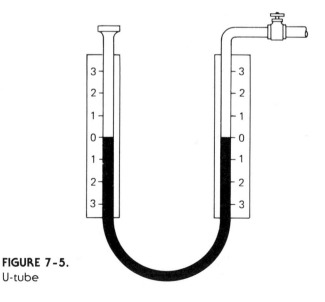

**FIGURE 7-5.**
U-tube

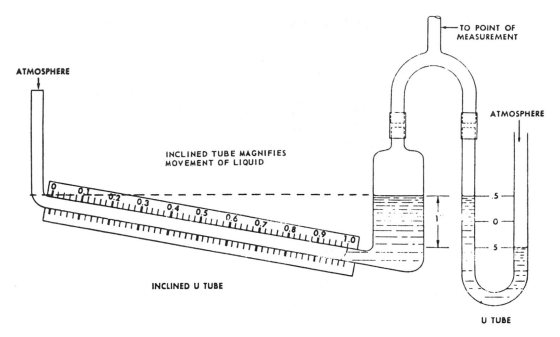

**FIGURE 7-6.**
Comparison of inclined-draft gauge and U-tube gauge

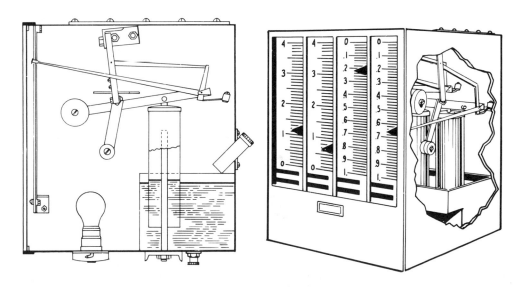

**FIGURE 7-7.**
Liquid-sealed draft gauge

as the U-tube, which is that the difference between the levels of the liquid in the two legs is a measure of the draft. (See Figure 7–7.)

The bottom of the inverted bell is sealed with oil or mercury, depending on the magnitude of the draft or pressure to be measured. It is supported by knife edges on the beam to reduce friction as much as possible. The weights counterbalance the weight of the bell, and the pointer is returned to zero. The source of draft is connected to the tube which projects into the inverted bell so that an increase in draft causes the pointer to move down.

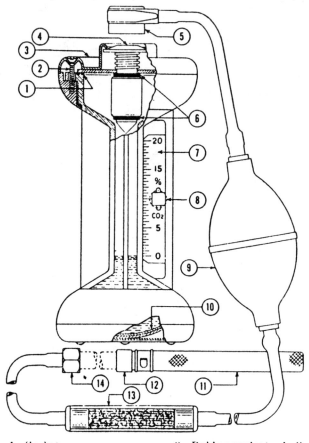

**FIGURE 7-8.**
$CO_2$ meter

| | | | |
|---|---|---|---|
| 1 | Gasket | 9 | Rubber aspirator bulb |
| 2 | Top cap holding screw | 10 | Bottom of analyzer |
| 3 | Top cap | 11 | Filter nipple |
| 4 | Plunger cap | 12 | Connection sampling tube to filter nipple |
| 5 | Connector tip | | |
| 6 | Plunger seats | 13 | Filter tube |
| 7 | $CO_2$ scale | 14 | Connection tubing to sampling tube |
| 8 | Scale locking screw | | |

## Measuring the $CO_2$

The carbon dioxide meter is used for determining, indicating, and recording the percentage of $CO_2$ in the products of combustion. (See Figure 7-8.) The principle this instrument is based on is that the specific weight of the flue gas varies in proportion to its $CO_2$ content. ($CO_2$ is considerably heavier than the remaining parts of the flue gas.)

# REVIEW QUESTIONS

1. What factors determine the amount of combustion air required by a boiler?
2. How is the air delivered to the boiler furnace in a forced-draft combustion system?
3. What type of draft system uses a blower in the boiler breeching?
4. What type of pressure is the boiler furnace under when an induced-draft system is used?
5. In what applications are combination forced-draft and induced-draft systems used?
6. What devices are used to connect the boiler to the chimney?
7. What conditions indicate that an automatic draft control be used?
8. Is it a good practice to install a barometric draft control in a room other than the boiler room?
9. Generally, to what pressure should barometric draft controls be set?
10. On what principle does the carbon dioxide ($CO_2$) meter operate?

# CHAPTER 8

# AUTOMATIC FUEL-BURNING EQUIPMENT

## Learning Objectives

The objectives of this chapter are to:

- Introduce the reader to the different types of automatic fuel-burning equipment
- Describe the various adjustments on burners needed to produce efficient combustion
- Show the different classifications of fuel-burning equipment
- Describe the different operating characteristics of the different types of burners.

Automatic fuel-burning equipment is a must for convenient, safe, and efficient operation of modern hot water and steam boiler applications. There are three basic types of automatic equipment used for burning fuel for boiler operation. They are categorized as (1) gas, (2) oil, and (3) solid fuel-burning equipment. In most applications more than one type of fuel can be used by using different types of fuel burning equipment. One type of fuel will be the main type and the other will be the standby type. (Examples are gas-oil and oil-gas.) It is the responsibility of the boiler operator to know how each type of burner operates, how to switch fuels, and how to make certain that the burner train is operating as efficiently as is possible.

# GAS-BURNING EQUIPMENT

A gas burner may be defined as a device that does the final mixing of the gas and air and delivers this mixture to the combustion zone. The two major types of gas burners are atmospheric injection and power burners. The gas may be ignited by a gas pilot, either standing or intermittent, or by an automatic spark ignitor, depending upon the manufacturer's design.

## Atmospheric Injection Gas Burners

Atmospheric injection gas burners are divided into drilled port, slotted port, and single port-type burners. These classifications are derived from the burner head ports. (See Figure 8-1.) They may be either an inshot type of burner or the up-shot type, depending upon the design of the equipment. (See Figure 8-2.)

Each burner is equipped with its own orifice, face, venturi, adjustment, and head. The orifice is the component that admits the correct amount of gas into the burner. The face is the place where the primary air is drawn into the burner. It usually has some type of adjustment for controlling the primary air that enters the burner. The gas and air is mixed in the venturi and the mixture picks up some turbulence and velocity here. The head is the part of the burner that is placed inside the combustion zone and admits the gas-air mixture into the combustion zone.

In atmospheric injection gas burners the secondary air is nonadjustable. Therefore, the $CO_2$ in the flue gases is nonadjustable. The equipment manufacturer establishes the amount of secondary air that can enter the combustion zone and this amount is noted by the American Gas Associa-

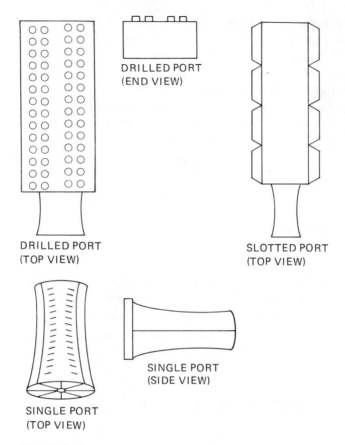

DRILLED PORT
(END VIEW)

DRILLED PORT
(TOP VIEW)

SLOTTED PORT
(TOP VIEW)

SINGLE PORT
(SIDE VIEW)

SINGLE PORT
(TOP VIEW)

**FIGURE 8-1.**
Types of atmospheric injection burners

tion. It should not be changed because the safe operation of the equipment may be jeopardized.

**Gas Burner Adjustment.**   The adjustment of gas burners is primarily a matter of adjusting the amount of primary air supplied to the burner. Before the air can be adjusted, the gas pressure must be adjusted to the proper pressure for the type of gas being used. Then the primary air supply to the burner should be closed off until yellow tips appear on the flame. More air is then slowly admitted into the burner until the yellow tips just disappear, then one-eighth turn more. This type of adjustment provides a ready ignition from port to port and a more quiet flame extinction.

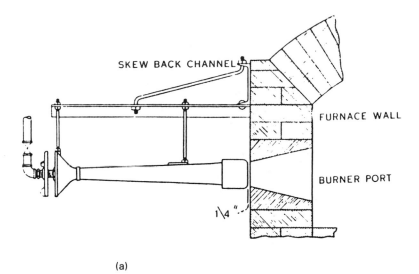

(a)

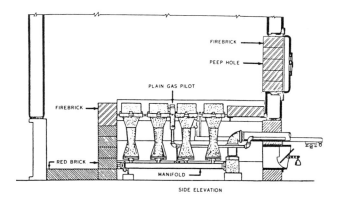

(b)

**FIGURE 8-2.**
(a) Inshot gas burner; (b) up-shot gas burner

## Power Gas Burners

These types of gas burners use a fan to deliver and mix the combustion air with the gas. They may be either a natural draft or a forced draft type of burner. (See Figure 8–3.) When the natural draft type is used a vent sys-

**FIGURE 8-3.**
Power gas burner (Courtesy of Ray Burner Co.)

tem is required to draw the flue gases out of the boiler; the fan provides only enough power to move the air through the burner. Most of them look much like an inshot atmospheric-type burner with a fan and windbox added. The combustion is more complete because of the more complex gas-air mixing patterns.

The use of power burners has created the need for larger, more powerful blowers. The fan moves the air through the burner as well as forcing the flue gases up the vent system. The vent system on these types of equipment functions merely to convey the flue gases from the equipment to the atmosphere. When a forced-draft power burner is used, a gas-tight boiler and vent system is required to prevent entrance of the products of combustion into the building.

With the use of power burners, superior control of the combustion process is accomplished, especially in restricted equipment. The high combustion efficiencies are obtained through a wide variety of loads because of the air-fuel mixture, and matching the flame pattern to the boiler furnace.

The boiler furnace and the heat transfer surfaces are designed for maximum heat transfer when they are used with the designed fuel. It is necessary to determine the maximum and the minimum gas pressure that should be applied to these burners so that the proper type of control system can be used for safety and efficiency.

**Operation.**   On demand from the control system, the gas is introduced into a controlled air stream, causing a thorough mixing of the gas and air. This mixing helps to maintain a stable flame front. The mixture then flows to the front of the burner where the ignition device is located. The gas-air mixture is ignited by the ignition system and the flame is directed into the combustion zone.

When a ring type of burner is used, the gas is introduced into the air stream through a gas-filled ring ahead of the combustion zone. In a premix type of burner, the gas and primary air are mixed together and then the mixture is mixed with the secondary air in the combustion zone.

## Adjustments

The adjustments on power type burners are mainly to the primary air. In most cases the secondary air is provided by natural draft which is not usually adjustable.

**Shutter Adjustment.**   When adjusting this type of burner, close off the air shutter until a yellow tip just appears on the flame. Then open the shutter until the yellow tips just disappear, then open the shutter one-eighth turn more. This type of flame provides ready ignition and a quiet extinction. It is always a good practice to check the amount of carbon monoxide in the flue gases. This is done with a suitable indicator. There should be no more than 0.04% carbon monoxide in an air free basis.

## Altitude Compensation

Most commercial and industrial units use forced-draft systems. In these systems the burner head, boiler combustion zone, and vent system act as orifices downstream of the combustion air fan. When they are used at higher altitudes larger burner fans are sometimes needed to increase the volume of air at a pressure great enough to overcome these restrictions.

# OIL-BURNING EQUIPMENT

An oil burner can be defined as a mechanical device used for preparing fuel oil to combine with air under controlled conditions for proper combustion. There are two methods used for preparing oil to be burned: atomization, and vaporization. Oil burners may be the natural draft type or the forced-draft type. The fuel oil may be ignited by one of three methods: (1) electric spark, (2) gas pilot flame, or (3) oil pilot flame. Fuel oil burners operate with either a luminous flame or a nonluminous flame. They may be operated continuously, intermittently, with a modulating control system, or with a high-low flame.

## Classification

Oil burners are classified by the type of application, type of atomization, or firing rate. They can be further broken down by type of design and operation as follows: pressure atomizing, air or steam atomizing, rotary, portable, vaporizing, and mechanical atomizing. All of them are capable of almost completely burning all of the fuel while producing almost no smoke when operated with 20% excess air. This amount of excess air relates to about 12% $CO_2$ in the flue gases.

**Pressure-Atomizing Oil Burner.**   This type of fuel oil burner is used in most installations in which number 2 grade fuel oil is used. (See Figure 8–4.) The oil is atomized with pressures ranging from 100 psig to 300 psig through the burner nozzle, which breaks the oil into a fine swirling mist as a cone-shaped spray and directs it into the combustion zone. (See Figure 8–5.) The combustion air fan forces the air through the burner and into the oil spray where the two are mixed in the proper proportions. The mixture is then ignited by either an electric spark, or a spark-ignited gas or oil pilot burner. (See Figure 8–6).

Pressure-atomizing oil burners may be either forced-draft or natural draft burners. The forced-draft type is equipped with a fan and motor with the capacity to supply all of the combustion air to the combustion zone with enough pressure to force the flue gases through the boiler, the vent system, and out into the atmosphere. It does this without any assistance from an induced-draft fan or from the draft caused by a chimney. The mixing of the oil and air under pressure reduces the amount of refractory needed to sup-

**FIGURE 8-4.**
A high-pressure oil burner (Courtesy of The Carlin Co.)

port the combustion process. The induced-draft type of oil burner operates with a negative pressure inside the boiler combustion zone.

The firing rate of pressure-atomizing oil burners can be changed by varying both the oil pressure and the air delivery to the burner nozzle. The air is generally regulated by shifting a damper in the air stream. This range

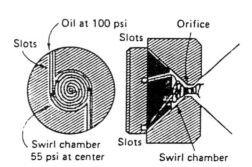

**FIGURE 8-5.**
Nozzle operation (Courtesy of Delavan Inc.)

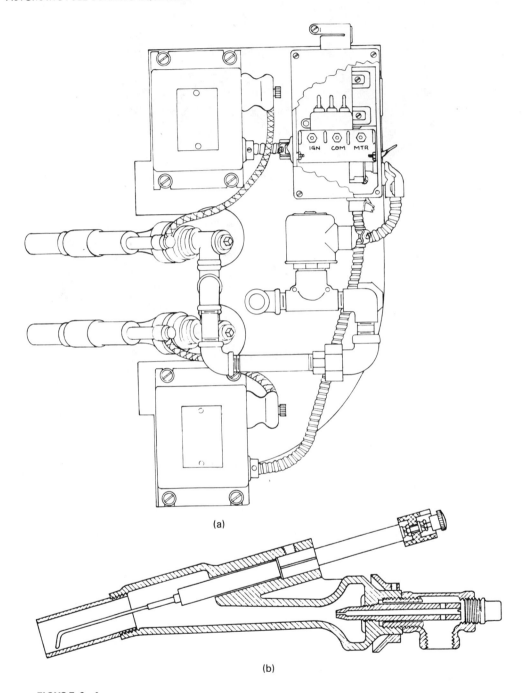

(a)

(b)

**FIGURE 8-6.**
Oil burner ignitor assemblies. (a) Ignitor bracket assembly; (b) raw gas ignitor. (Courtesy of Ray Burner Co.)

adjustment is usually limited to about 1.6 to 1 for any given nozzle size. The firing mode controls will vary according to the burner size and the desires of the burner manufacturer. Most of the larger burners are equipped with controls which will provide the desired variation in the heat input to the equipment.

When using a number 2 fuel oil in these burners, no preheating of the fuel is necessary. However, number 4 fuel oil must be preheated to about 100°F for proper atomization and burning. These burners, when properly adjusted, will operate well with 20% excess air, which provides approximately 12% $CO_2$, with no visible smoke and only a trace of carbon monoxide in the flue gases.

Good operation of pressure-atomizing oil burners depends upon a relatively constant draft in either the combustion zone or at the breeching connection, depending upon which burner is selected.

**Return-Flow Pressure-Atomizing Oil Burner.** Return-flow pressure atomizing oil burners are very close in operation and design to the pressure-atomizing type of oil burners. They are sometimes called modulating pressure-atomizing oil burners. They can be operated over a wide load range with approximately a 3 to 1 reduction in the firing rate for any specific atomizer.

The use of the return flow nozzle makes this wide range possible. It has an atomizing swirl chamber located just upstream of the orifice. A high rate of oil flow and a high pressure drop through the swirl chamber is maintained throughout the complete variation in the load and firing range. This high oil flow rate and pressure drop provides the good atomization, which is responsible for the efficiency of the burner at low-firing rates. Any excess oil pumped over the demand either returns to the oil storage tank, or passes to the suction inlet of the fuel pump from the swirl chamber.

The variation in the firing rate is obtained by varying the oil pressure in both the inlet of the oil pump and in the oil return lines.

The operation of this type of oil burner is, with the exception of the atomizer, the same as the pressure-atomizing oil burner.

**Air-Atomizing Oil Burners.** This burner is very much like the pressure-atomizing burner with the exception of the burner nozzle. The air required for atomization, and the oil, are delivered to different parts of the burner nozzle. The nozzle is designed so that air will break the oil into very fine droplets. This atomized oil and air mixture is then forced into the combustion zone through the outlet orifice by the velocity of the air. (See Figure 8-7.)

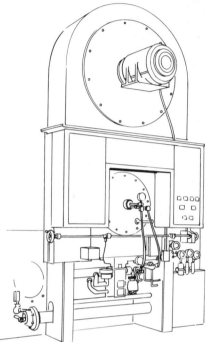

**FIGURE 8-7.**
Air-atomizing burner
(Courtesy of Cleaver-Brooks)

The air for combustion is forced by the combustion air fan through the throat of the burner. It then is mixed thoroughly with the oil and air mixture from the nozzle inside the combustion zone. The fuel is then ignited with an ignitor very similar to the ones discussed previously.

Air atomizing burners can use heavy fuel oils such as the number 6 grade. Number 2 fuel oil requires no preheating, but the heavier oils do, so that an adequate viscosity can be maintained for proper atomization. They also have a wide variety of capacities without changing the nozzle size. The smaller sizes are capable of a turn down of about 3 to 1 and the larger sizes are capable of 6 or 8 to 1. These variations are achieved by changing the amount of combustion air, the oil pressure, and the atomizing air that is delivered to the burner. Some designs of this type of burner require as low as 5 psig atomizing air pressure and other designs may use up to 75 psig.

This type of burner requires from 2.2 to 7.7 ft$^3$ of compressed air per gallon of fuel oil for proper operation. Air-atomizing burners can operate with a wide load range and are therefore very well adapted to the modulating type of control system.

Air-atomizing oil burners operate well with 15 to 25% excess air, which provides approximately 12 to 14% $CO_2$ when operated under full load conditions. Under these conditions there will be no visible smoke (a number 2 smoke spot on the smoke test gauge) with only a slight trace of carbon monoxide in the flue gases. (A smoke test gauge has holes which are covered with smoke-colored glass. Each hole is a different shade. When the smoke from a vent can be seen through a given hole it has a higher smoke value.)

**Horizontal Rotary Cup Oil Burners.** Horizontal rotary cup oil burners atomize the oil by spinning it off the edge of a cup rotating at high speed. The oil leaves the cup in a very thin film which is then mixed with the high-velocity primary air. The mixture then passes through an annular nozzle which surrounds the outer rim of the atomizing cup.

In most units, the atomizing cup and the primary air fan are positioned on a horizontal shaft which is connected to the driving motor. The shaft, cup, and fan rotate at a constant speed. The speed is determined by the make and the size of the burner. The oil pump is driven by the horizontal shaft through the use of a worm and gear train. The oil is fed to the atomizing cup at a controlled rate in accordance with the unit design characteristics. (See Figure 8–8.)

The secondary air is delivered through the burner windbox by its own fan. The fuel oil is ignited by a spark-ignited gas or oil pilot burner. Horizontal rotary cup oil burners are very well suited for use with modulating-type controls because of their wide load range of about 4 to 1.

These types of burners, when properly adjusted, operate very satisfactorily with 20 to 25% excess air, which provides approximately 12.5 to 12% $CO_2$ in the flue gases at full load conditions. They will produce no visible smoke (about a number 2 smoke spot) and only a trace of carbon monoxide in the flue gases.

Horizontal rotary cup burners require a relatively constant source of draft to the combustion zone. These burners are mounted on the front of the boiler with a swingaway mount for convenience in service and maintenance procedures.

**Steam-Atomizing Oil Burners.** Steam-atomizing oil burners use the impact and expansion of steam for atomization of the oil. The oil and steam flow through the burner gun and to the burner nozzle through separate channels.

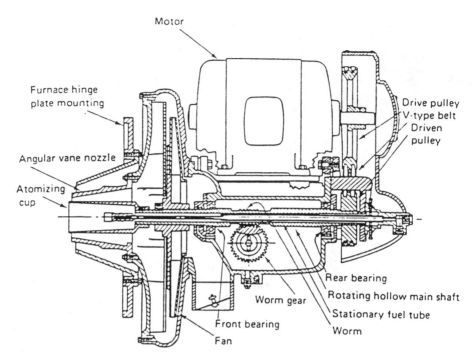

**FIGURE 8-8.**
Components of a Ray rotary oil burner (Courtesy of Ray Burner Co.)

The oil and steam are mixed before they are discharged through the nozzle orifice into the combustion zone.

A forced-draft fan forces the combustion air through the directing vanes in the burner register. The vanes give the air a spinning motion as it enters the burner throat, which directs it into the cone-shaped oil spray. The oil and air are thoroughly mixed at this point and then the mixture enters the combustion zone. (See Figure 8–9.)

These burners generally operate with an oil pressure of about 100 to 150 psig at full load conditions. In most burners, the steam pressure is maintained at about 25 psig greater than the oil pressure. The capacity of this type of burner is adjusted by varying the oil pressure to the burner nozzle. Because of their variable capacity, steam atomizing oil burners are adaptable to modulating control systems.

Some of these types of burners are designed to operate with air pressure for atomizing the oil on boiler start-up; others are designed so that a pressure-atomizing nozzle can be used when neither steam or air pressure is available. The burners are switched over to steam atomization when the steam

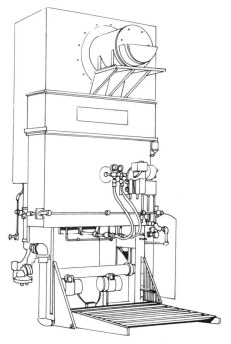

**FIGURE 8-9.**
Steam-atomizing burner
(Courtesy of Cleaver-Brooks)

pressure has risen sufficiently to permit its use. Steam-atomizing burners use about 1 to 5 lb of steam per gallon of fuel oil used.

Steam-atomizing burners require that oils heavier than number 2 be preheated to the proper viscosity so that the best atomization can be obtained. These burners operate satisfactorily with about 15% excess air which provides about 14% $CO_2$ in the flue gases at full load conditions. When properly adjusted, these burners operate with no visible smoke and have only a trace of carbon monoxide in the flue gases.

**Mechanical-Atomizing Oil Burners.**    These types of burners atomize the oil by pressure of the oil alone. The oil is atomized when it passes through the nozzle at the pressure supplied by the pump.

Mechanical-atomizing burners are usually equipped with an air register, or windbox. The register has a series of adjustable vanes. The combustion air fan is usually mounted separately from the burner, and the air is forced through a duct to the register.

The capacity of these burners is obtained by varying the oil pressure from the pump. The oil pressure varies from about 90 to 900 psig. This type of burner has a narrow load range.

# SOLID FUEL-BURNING EQUIPMENT

Solid fuel-burning equipment (often referred to as stokers) is defined as a mechanical device used for feeding a solid fuel into the combustion zone of a boiler. This device automatically provides the combustion air for burning the fuel, and it may have some means for automatically removing the ashes. In most instances the type of fuel used is coal.

## Stoker Types

The introduction of stokers permitted an increase in the efficiency of coal as a boiler fuel. Through their use more coal could be burned in a small boiler and larger boilers could be designed because of the increased feeding capabilities.

Stokers are designed to evenly feed the coal into the furnace of the boiler. Their use eliminates the need to open the furnace door to feed fuel into the combustion zone of the boiler, which allowed the firebricks to cool down and chill the boiler around the door. The size of the boiler usually determines the number of stokers used for each plant size.

The stoker has various components, such as a coal hopper, a method of feeding, a fan, and a removal system. The hopper is for holding the coal that is to be fed into the boiler furnace. The feeding mechanism is for moving the coal from the hopper into the furnace, and the fan is used to supply combustion air to the coal bed.

There are basically four types of stokers in use which are classified by the type of feeding system used. They are: (1) underfeed, (2) vibrating grate, (3) chain grate or traveling grate, and (4) spreader.

**Underfeed Stokers.**  The underfeed stokers feed the coal into the retort which is located beneath the burning bed of coal. Underfeed stokers are classified as gravity feed and horizontal feed types. When the gravity feed types are used the coal is fed into a retort which is designed on an incline of about 25°. The gravity feed types may have multiple retorts. When the horizontal feed types are used the coal is fed into a retort which is parallel with the floor. Horizontal feed stokers are usually designed to use either one or two retorts, but seldom are more than two used.

*Gravity Feed Stokers.*    Gravity underfeed stokers are equipped with a series of sloping multiple retorts. The coal is fed into each retort and is moved slowly to the rear of the furnace while it is being forced in an upward direction over the retorts. When the coal is completely burned the ashes are removed through an ash discharge in the rear of the boiler.

*Horizontal Feed Stoker.*    When the horizontal type of stoker is used the coal is fed into the retort by one of two methods: screw feed, or ram feed.

The screw feed method is mostly used on the smaller size boilers while the ram feed method is used on the larger capacity units. After the retort is full, the coal is pushed up and over the retort where the fuel bed is formed. Tuyers are located on each side of the retort. The combustion air is supplied through these tuyers and through the air ports located on the sides of the grates. (See Figure 8–10.) Additional combustion air is supplied directly to the flame zone through the overfire air ports which are located above the fuel bed. The overfire air is used to prevent smoking, especially during low load conditions.

**Vibrating-Grate Stoker.**    The vibrating-grate type of stoker is an overfeed, mass-burning, continuous ash-discharge unit. (See Figure 8–11.) In this type of stoker, the vibrating grate is supported by vertical plates that are evenly spaced and move with a back and forth motion. This motion causes the fuel to pass through an adjustable gate and into the combustion zone of the boiler. The combustion air is forced into the combustion zone through the openings created by the movement of the grate supporting plates. The grates are connected to the boiler-water circulating system and are cooled by the flow of water. As the ash is forced through the retort, it is automatically discharged into an ash pit. It is then removed and discarded.

Vibrating-grate stokers are very adaptable to the different types of coal that are available. They are simple in operation, require a low maintenance schedule, and have a wide turn-down ratio. The coal is fed into the retort, and the movement of the fuel bed is regulated in response to the frequency and duration of the vibration of the plates. The coal-burning rate is regu-

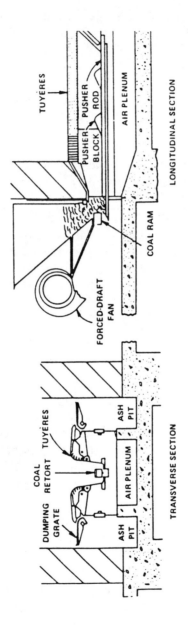

**FIGURE 8-10.**
Horizontal underfeed stoker with single retort (Courtesy of ASHRAE)

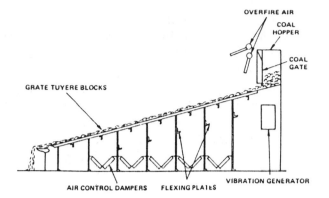

**FIGURE 8-11.**
Vibrating-grate stoker
(Courtesy of ASHRAE)

lated by automatic controls which govern the amount of combustion air being supplied to the furnace.

**Chain-Grate or Traveling-Grate Stokers.**   Chain-grate or traveling-grate stokers are sometimes used interchangeably because their methods of operation are very similar. The only major difference is in the design of the grate. (See Figure 8–12.)

In the chain-grate stokers the chain links move with a scissorlike motion at the return bend of the stoker. In the traveling-grate stoker there is almost no movement between the two grate sections. When coals which have clinkering-ash characteristics are to be used, the chain grate type is preferred over the traveling grate type.

When these types of stokers are to be used, both front and rear arches are available which will improve the combustion characteristics by reflecting heat back into the fuel bed. In operation, the front arch helps to mix the

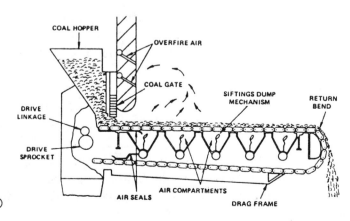

**FIGURE 8-12.**
Chain-grate stoker
(Courtesy of ASHRAE)

stream of rich volatile gases with the combustion air, which improves the combustion rate by reducing the amount of unburned hydrocarbons in the flue gases. In later designs, overfire air ports replaced the front arch which was used for mixing the volatiles and the unburned hydrocarbons. Each zone of the grate is equipped with its own individual air damper which is used to control the pressure and volume of the combustion air being supplied to that section of the grate.

During the operation of both of these types of stokers, the coal is fed from the hopper onto a moving grate. The coal then passes through the adjustable grate which controls the thickness of the coal bed. It then passes into the boiler furnace where it is preheated by the heat radiating from the refractory bricks or by the furnace gases. The volatile gases that are driven from the coal by this heat are then ignited. The coal continues to burn as it is moved along by the moving grate. The fuel bed gets thinner and thinner until the combustion process is completed at the rear of the grate travel. At this point the ash is dumped into the ash pit, as the grates turn downward to return to the other end of the boiler.

**Spreader-Type Stokers.** Spreader-type stokers make use of both the suspension and the grate types of methods of burning fuel. This type of stoker is designed to burn about one-half of the coal in suspension; the remainder is burned on the grate. This procedure causes a much higher particulate loading than the other types of stokers. Because of this particulate matter, a dust collector is required to trap these particles in the flue gases before they are allowed to flow into the vent stack. Fly-carbon reinjection systems are used to return these carbon particles to the furnace of the boiler for reburning. This process requires that a very high-efficiency dust collector be used.

In operation, the coal is continually fed into the boiler furnace above the already burning bed of coal. (See Figure 8–13.) While in suspension, the coal fines are partially burned. The large particles fall into a thin, fast-burning fuel bed on the grate.

The spreader stoker is the most popular type in industrial installations for two reasons.

1. There is an almost instantaneous firing rate of the fuel.
2. This type of firing is very flexible in response to load fluctuations.

There are several different types of grates used in spreader-type stokers and all of them are designed to provide a high resistance to the flow of combustion air. This is necessary to prevent blowholes in the thin fuel bed.

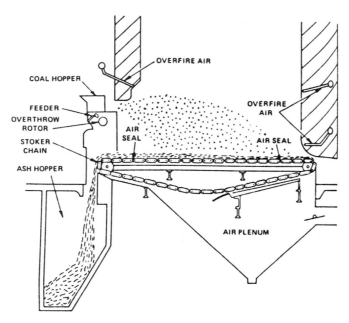

**FIGURE 8-13.**
Spreader stoker, traveling-grate type (Courtesy of ASHRAE)

In the smaller types the grates are sectionalized and the ashes are manually removed. The sections permit undergrate air chambers for each section of the spreader unit. Therefore, both the fuel supply and the air supply can be temporarily stopped to allow cleaning and maintenance without stopping the complete boiler operation.

In later, larger models of spreader-type stokers, a continuous ash-discharge type of grate is used. The combining of the spreader-type stoker with the traveling grate permits an increase in the burning rate of the fuel of about 70%.

# REVIEW QUESTIONS

1. Name the two major types of gas burners.
2. What is the name of the device that admits gas to a gas burner?
3. What does the adjustment of a gas burner control?
4. What has the use of power gas burners accomplished?

5. Why is it necessary to determine the maximum and minimum gas pressures to be used with a power gas burner?

6. On a power gas burner, to what are most of the adjustments made?

7. Name the two methods of preparing fuel oil to be burned.

8. Name the three methods used on boilers to ignite the oil.

9. What is the most popular grade of fuel oil used in pressure atomizing oil burners?

10. How can the firing rate of a pressure-atomizing oil burner be changed?

11. In the pressure-atomizing type of oil burner, to what temperature should number 4 fuel oil be heated?

12. How is the firing rate changed with the return-flow pressure-atomizing oil burner?

13. What is the difference between pressure-atomizing and air-atomizing oil burners?

14. What force is used to atomize the oil in mechanical-atomizing oil burners?

15. What is the purpose of the stoker on coal burning equipment?

16. Name the four types of stokers used.

17. On coal-burning equipment, how is combustion air fed directly to the flame zone?

18. In the vibrating-grate stoker, what causes the coal to flow into the furnace?

19. In what type of stoker is about one-half of the coal burned in suspension and the rest burned on the grate?

# CHAPTER 9

# FEED WATER CONTROLS

## Learning Objectives

The objectives of this chapter are to:

- Introduce the reader to the basic feed water system controls used on boiler installations
- Show the uses of the different components used in the feed water system of a boiler
- Delineate a typical feed water system
- Describe the operation of feed water system controls.

When a boiler is producing steam, some of the water is lost through leaks and cannot be recovered. Most of the water that is used for process purposes can be recovered through the condensate return system. The purpose of the feed water system and controls is to replenish this lost water and to return the condensate to the boiler so that it can operate efficiently and safely. Some installations require that a deaerating tank be used in conjunction with a condensate receiver; others operate satisfactorily with only the condensate receiver.

Some of the more complex systems also require that a vacuum pump (condensate return pump) be used along with the condensate receiver. The purpose of the vacuum pump is to aid in the return of the condensate to the boiler during operation.

# CONDENSATE TANK

The purpose of the condensate tank is to store the condensate return water until it is needed by the boiler; to allow city make-up water to enter the system; and to heat the feed water to the proper temperature. If a deaerater is incorporated, it provides the deaeration. (This is described in Chapter 10 as a separate function.) The water content of the condensate tank must be about one-sixth of the maximum evaporation capacity of the boiler or boilers to which it is connected. This amount of water will cover about 10 minutes of normal operation.

When the volume of a system is above average, or the returns are slow to come back to the tank, or when the feed water needs to be heated before entering the boiler, the tank must have a larger volume which will cover about 20 to 30 minutes of normal operation.

When the feed water is below 180°F, it must be heated to conserve energy, to prevent the boiler from cooling down during the water make-up process, and to help eliminate thermal shock. The heater capacity must be increased proportionally to the percentage of make-up water.

# FEED WATER SYSTEM AND CONTROLS

The following is a description of the controls which are used in a typical feed water system. Refer to the diagram in Figure 9–1 for the following discussion.

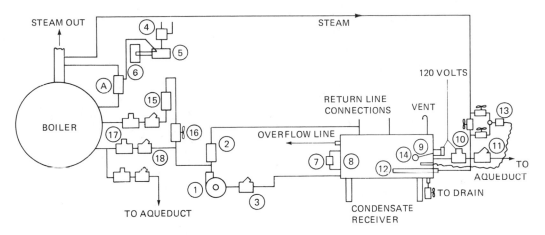

**FIGURE 9-1.**
Boiler feed water system

## Water Level Control

The water level control is installed on the boiler at the desired water level (A). (See Figure 9–2 also.) It generally consists of a float-operated set of contacts. When the water level inside the boiler drops about 3/4 in. below the normal operating level, the electrical contacts close and start the condensate pump (1). The pump then transfers water from the condensate receiver into the boiler. When the water in the boiler reaches its normal level, the float in the water level control will raise and open the electrical contacts, stopping the pump motor.

## Condensate Strainer

The strainer (3) is installed between the condensate receiver and the pump turbine. (Figure 9–3.) Its purpose is to trap foreign particles in the water and prevent them from entering the pump turbine and causing possible damage.

## Relief Valve

The relief valve (2) is installed on the outlet of the pump with a line going back into the condensate receiver. (Figure 9–4.) Its purpose is to protect the pump against any excessive pressures. If the pump discharge pres-

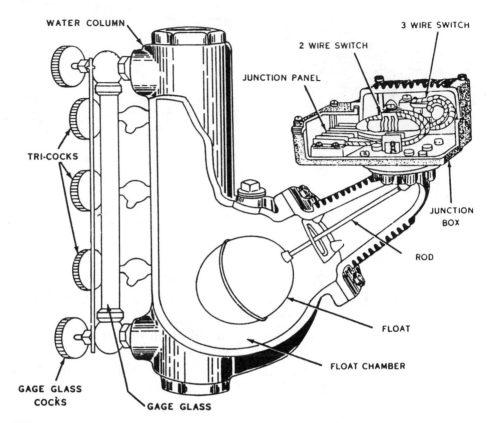

**FIGURE 9-2.**
Cross section of water feeder cutoff combination

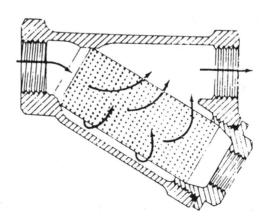

**FIGURE 9-3.**
Strainer cutaway

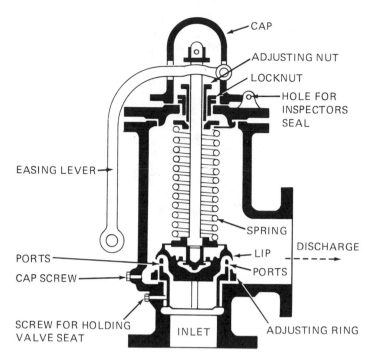

CAP

ADJUSTING NUT

LOCKNUT

HOLE FOR INSPECTORS SEAL

EASING LEVER

SPRING

PORTS

CAP SCREW

LIP

DISCHARGE

PORTS

**FIGURE 9-4.**
Relief valve cutaway

SCREW FOR HOLDING VALVE SEAT

INLET

ADJUSTING RING

sures should reach a dangerous point, the relief valve will open and allow the treated water to return to the condensate receiver rather than wasting it down the sewer or overloading the pump motor. If this condition should exist it must be checked out and the problem must be corrected or the boiler could run short of water and overheat.

## Check Valve

As the feed water leaves the pump, it will be divided in two directions. Most of it will pass through a check valve before it can enter the boiler (18). (See Figure 9–5.) The purpose of the check valve is to allow the water to enter the boiler and prevent it from backing up into the condensate receiver, which would leave the boiler low on water. When the pump is running, the check valve will open and permit the water to pass. When the pump is not running or the boiler pressure exceeds the pump pressure, the check valve will close, stopping the flow of water.

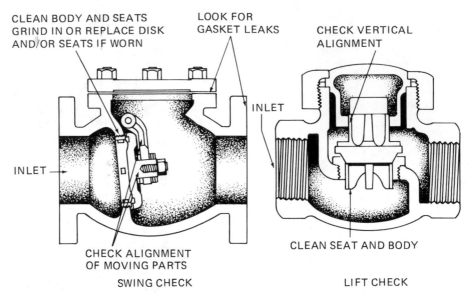

**FIGURE 9-5.**
Check valves

## Stop Valve

After leaving the check valve, the feed water will pass through a stop valve (17). The purpose of the stop valve is to allow the check valve to be isolated for repairs without draining down the boiler. The stop valve is normally a globe valve that is manually operated. To stop the flow of water, turn the valve stem in. When the work is completed the stem is turned out to allow the water to flow into the boiler. During normal operation this valve is in the open position.

## Manual Shot (Chemical Pot) Feeder

When a chemical pot feeder is used, a portion of the feed water passes through it before going into the boiler (15). (See Figure 9–6.) The purpose of the chemical pot feeder is to introduce chemicals into the boiler on a controlled basis. There are also two stop valves and a check valve in this line. The purpose of the check valve is prevent the back-flow of water

**FIGURE 9-6.**
Manual shot (chemical pot) feeder (Courtesy of Cleaver-Brooks)

through the feeder. The two stop valves are used to close off the pot feeder so that chemicals can be placed in it, and they are used to regulate the flow of water so that the proper amount of chemicals can be added to the boiler feed water.

## Condensate Receiver

The purpose of the condensate receiver is to

1. Collect the condensate after it has passed through the process equipment
2. Take on fresh make-up water
3. Heat the water to the desired temperature.

The feed water will remain in the condensate receiver until the boiler calls for more water; then the pump will transfer the water from the tank to the boiler.

## Temperature Regulator

The temperature regulator is installed in a steam line to the condensate receiver (13). The purpose of the temperature regulator is to allow steam to pass to the water heater and heat the feed water to the desired temperature. It is usually installed with three stop valves and a strainer. Two of the valves are used to isolate the temperature regulator for repairs. The other stop valve is used to manually feed steam to the heater to maintain the feed water temperature.

## Temperature Sensing Element

The temperature sensing element is installed inside the condensate receiver (14). The purpose of the sensing element is to signal the temperature regulator when either more or less heat is needed.

## Steam Water Heater

The steam water heater is located inside the condensate receiver (12). The purpose of the heater is to receive the steam from the temperature regulator and direct it into the feed water so that it can be heated to the desired temperature.

## Water Gauge Glass

The water gauge glass is located on the condensate receiver at the proper water level (7). (See Figure 9–7.) The purpose of the water gauge glass is to help the boiler technician in determining what the water level is inside the condensate receiver.

## Thermometer

A thermometer is located on the condensate receiver to indicate the temperature of the water inside the tank (8). The boiler technician can determine if the temperature is at the proper level just by looking at the thermometer.

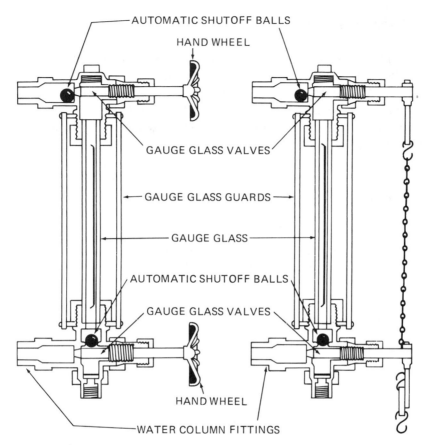

**FIGURE 9-7.**
Cross sections of typical gauge glasses

## Float Switch

The float switch is located inside the condensate receiver (9). The purpose of the float switch is to monitor the feed water level inside the tank and signal the fresh water solenoid valve when make-up water is needed. When the water level falls, the switch will cause the water solenoid to open and let water into the tank. When the water level has risen to the proper level the switch will cause the water solenoid to close, and thus stop the flow of water.

## Solenoid Valve

The freshwater solenoid valve is located in the city water line to the condensate receiver (10). The purpose of this valve is to admit city water into the condensate receiver when more water is needed than is being returned from the condensate system. When the float switch inside the receiver determines that more water is needed, it will signal the solenoid valve to open and allow the correct amount of water to enter the system.

## Aqueduct Strainer

The aqueduct strainer is located in the city water make-up line to the system (11). The purpose of this strainer is to remove foreign particles from the water before it enters the feed water system.

## Fusible Electric Switch

This switch is located in the electric line to the pump (4). Its functions are (1) to provide protection to the electrical circuit, and (2) to permit turning off the electricity to the pump if repairs are needed. At times the fuses in this switch will blow and will need to be replaced. However, when this condition occurs be sure to correct the problem before replacing the fuses to prevent a recurrence of the problem.

## Magnetic Starter

The magnetic starter is installed in the electrical line to the pump after the fusible disconnect switch (5). The purpose of the magnetic starter is to start and stop the pump motor in response to the demands from the water level control.

## Transformer

The transformer is installed in the electrical line between the line voltage circuit and the low voltage circuit (6). The purpose of the transformer is to provide low voltage to the control circuits that are used to operate some of the electrical control circuits used in the operation of the boiler.

## Vacuum Pump

A vacuum pump is used on steam systems that have a large number of condensate return lines, or when condensate return is a problem. The purpose of the vacuum pump is to aid in the return of the condensate to the boiler. The condensate is pumped from the vacuum pump receiver tank and is discharged into the condensate receiver tank. The condensate pump then pumps the condensate into the boiler in the normal manner. Vacuum pump motors are controlled by a float switch which signals the pump that the tank is full of water and needs to be emptied.

## REVIEW QUESTIONS

1. Where is condensate return water stored?
2. When is it necessary to heat feed water?
3. Where is the water level control installed on a boiler?
4. How is the condensate pump protected from excessive pressures?
5. What is the purpose of the check valve?
6. How are boiler system components isolated for repairs?
7. How are chemicals sometimes fed into a boiler?
8. How is the water level inside a condensate tank determined?
9. What device determines when make-up water is needed in the boiler?
10. What is the purpose of strainers used in water systems?

# CHAPTER 10

# WATER TREATMENT

## Learning Objectives

The objectives of this chapter are to:

- Show the need for boiler water treatment
- Introduce the reader to the different components of boiler feed water
- Demonstrate the effects of each boiler feed water component
- Describe different procedures used in boiler feed water testing
- Make the reader aware of the safety procedures used when testing boiler feed water.

All natural waters contain acid materials and the scale-forming compounds of calcium and magnesium, which attack ferrous materials. Some waters contain more scale-forming compounds than others, and some waters are more corrosive than others. Subsurface or well waters are generally more scale-forming, while surface waters are usually corrosive. To prevent scale formation on the internal water-contacted surfaces and to prevent damage to the boiler metal by corrosion, feed and boiler water must be chemically treated. This chemical treatment prolongs the useful life of the boiler and results in appreciable savings in fuel, since maximum heat transfer is possible when no scale deposits occur.

# SCALE

A crystal clear water, satisfactory for domestic use, may contain enough scale-forming elements to render it harmful and dangerous in boilers. Two such scale-forming elements are precipitates of hardness, and silica.

Scale deposited on the metal surfaces of boilers and auxiliary water heat exchange equipment consists largely of precipitates of hardness—the ingredients calcium and magnesium, and their compounds. Calcium sulfate scale is, next to silica, the most adherent and difficult to remove. Calcium and magnesium carbonates are the most common. Their removal requires tedious hand-scraping and internal cleaning by power-driven wire brushes. If deposits are thick and hard, the more costly and hazardous method of inhibited acid cleaning is used. Scale deposits are prevented by (1) removing calcium and magnesium in feed water to boiler (external treatment); and, (2) chemically treating boiler water with agents such as phosphates and organic extracts to change scale-forming compounds to soft nonadherent sludge, which is easily removed from the boiler by boiler blowdown (internal treatment).

## Silica

Silica present in boiler feed water precipitates and forms a hard, glassy coating on the internal surfaces. In the feed water of high-pressure boilers, such as those used in electric generating plants, a certain amount of silica vaporizes under the influence of high pressure and temperature. The vapor is carried over with the steam, and silica is deposited on intermediate- and low-pressure blading of turbines. In boilers operating in the range of 20 to

125 psig, the silica problem is not so troublesome. If the water is low in hardness, contains phosphate that prevents calcium silicate scale from forming, or has sufficient alkalinity to keep the silica soluble, no great difficulty is encountered. The amount of soluble silica can be limited by continuous or routine boiler blowdown, which prevents buildup of excessive concentrations.

# CORROSION

Corrosion is concurrent with the problem of scale control. Boilers, feed water heaters, and associated piping must be protected against corrosion. Corrosion results from water that is acidic, which means it contains dissolved oxygen and carbon dioxide. Corrosion is prevented by deaerating the feed water to remove these dissolved gases; by neutralizing traces of dissolved gases in the outlet of the deaerating heater with chemicals; and by neutralizing acidity in the water with an alkali.

# METHODS OF TREATMENT

The specific method of chemical treatment employed will vary according to the type of boiler and the specific properties of the water from which the boiler feed is taken. In general, however, the chemical treatment of feed and boiler water is divided into two broad types or methods: external and internal treatment. External treatment consists of controlling alkalinity in the make-up water and removing scale-forming materials and dissolved gases (oxygen and carbon dioxide) before the water enters the boiler. Internal treatment entails the introduction of chemicals directly to the boiler feed water or the water inside the boiler. Frequently a combination of both external and internal chemical treatment is used.

External treatment, frequently followed by some internal treatment, can give better boiler water conditions than internal treatment alone. However, external treatment requires the use of a considerable amount of equipment such as chemical tanks, softening tanks, filters, or beds of minerals, and involves high installation costs. Such treatment is therefore used only where the make-up water available is so hard or so high in dissolved minerals that internal treatment by itself does not maintain the desired boiler water conditions, or where conditions of operation are such that internal treatment is not sufficient. The dividing line, where hardness and the concentration of dissolved mineral matter in the water are high enough so that external treatment must be used, depends upon certain mechanical considerations which relate to the particular plant. These considerations include the type and design of the boilers; the pressure and rating at which the boilers operate; the percentage of make-up water being used; the amount of sludge that can be tolerated in the boiler; and such general considerations as the space available and the adaptability of the operators.

Many methods of internal treatment are in use. Most of these treatments utilize carefully controlled boiler water alkalinity, an alkaline phosphate, and organic material. One of the organic materials used is tannin. Tannin is a boiler water sludge dispersant. That is, it makes solids in the water more fluid and prevents their jelling into masses that are difficult to remove by blowdown. Because of treatment costs and the simplicity of chemical concentration control, the alkaline phosphate-tannin method of internal treatment is perhaps the most widely used method. When properly applied and controlled, this treatment prevents the formation of scale on internal boiler surfaces and prevents corrosion of the boiler tubes and shell.

# BOILER WATER TESTS

As we have just seen, boiler water must be treated with chemicals to prevent the formation of scale on the internal surface of the boiler, and to prevent deterioration of the boiler metal by corrosion. Testing of the boiler water is necessary to determine if the amount of chemical residuals required to maintain clean boiler surfaces is present. As a boiler operator, you should be able to perform various kinds of boiler water tests. A few of the tests that you may be called upon to do are described below. Tests for hardness, phosphate, tannin, caustic alkalinity (with and without tannin), sodium sulfite, and pH are included. A test kit is provided for each test. The kit for a particular test will contain the equipment and materials required to make the test.

Before proceeding, here is a brief word of caution which applies to each test that will be discussed: *If the testing procedure of the equipment and/ or reagent supplier differs from that prescribed in this text, the supplier's procedure should be used.*

## Test for Hardness

Boilers operating at 15 psig or less are normally used for space heating and hot-water generation. Practically all the condensate is returned to the plant. Only a small amount of make-up water is required, and secondary feed water treatment usually is sufficient. When appreciable quantities of steam are used in process work and not returned as condensate to the plant, the problem of scaling and corrosion arises, and a more complete treatment of the feed water must be considered.

The ideal water for boilers is one that does not form scale or deposits; does not pit feed water system and boiler surfaces; and does not generate appreciable $CO_2$ in the steam. However, such a raw make-up water is impossible to obtain in the neutral state from wells or surface sources. It is necessary, therefore, to treat it to the extent that treatment can be justified economically in terms of advantage gained.

Feed water of 20 to 25 parts per million (ppm) of hardness as calcium carbonate ($CaCO_3$) need not be treated externally for reduction of hardness if:

1. Sufficient alkalinity is present to precipitate the hardness in the boiler as $CaCO_3$, or
2. Hardness reducers such as phosphates are introduced to combine with and precipitate the hardness.

Precipitation of this hardness in a low- or medium-pressure boiler generally will not cause wasteful blowdown. When the mixture of condensate and makeup in a medium-pressure steam plant is found to have a hardness greater than 20 to 25 ppm as $CaCO_3$, the hardness should be reduced to 0 to 2 ppm as $CaCO_3$.

Feed water of a hardness in excess of 2 ppm as $CaCO_3$ should be treated to bring it within the range of 0 to 2 ppm as $CaCO_3$. This small remaining hardness can be precipitated in the boiler by secondary treatment and removed by continuous blowoff equipment.

The test for hardness, as presented here, makes use of the Colorimetric Titration Method. This test is based on the determination of the total calcium and magnesium content of a sample by titration with a sequestering agent in the presence of an organic dye sensitive to calcium and magnesium ions. The end point is a color change from red to blue. It occurs when all the calcium and magnesium ions are separated.

The following equipment is used in making this test:

One 25-ml automatic, complete burette

One 210-ml porcelain casserole

One 50-ml graduated cylinder

One glass stirring rod

One calibrated dropper

The reagents are

Hardness indicator

Hardness buffer

Hardness titrating solution

To make the test, start by measuring 50 ml of the sample in the graduated cylinder and transfer it to the casserole. With the calibrated dropper, now add 0.5 ml of the hardness buffer reagent to the sample, and stir. Then add 4 to 6 drops of the hardness indicator. If hardness is present, the sample will turn red. Add the hardness titrating solution slowly from the burette, with continued stirring. When approaching the end point, the sample begins to show some blue coloration, but a definite reddish tinge can still be seen. The end point is the final discharge of the reddish tinge. The addition of more hardness titrate solution does not produce further color change.

In this procedure, the hardness titrating solution must be added slowly because the end point is sharp and rapid. For routine hardness determination it is suggested that 50 ml of the sample be measured, but only approximately 40 to 50 ml be added to the casserole at the start of the test. The hardness buffer reagent and the hardness indicator should then be added as directed, and the mixture titrated rapidly to the end point. The remaining portion of the sample should then be added. The hardness present in the remainder of the sample turns the contents of the casserole red again. Titrating is continued slowly until the final end point is reached. A record should be kept of the total milliliters of hardness titrating solution used.

## Test for Phosphate

This is a colorimetric test for phosphate, using a decolorizing carbon for removal of the tannin. Carbon absorbs the tannin, and the carbon and tannin are then filtered out. When tannin is not present, carbon improves the test for residual phosphate by making the tricalcium phosphate sludge more filterable.

The equipment required for the phosphate test includes

One phosphate color comparator block of two standards—30 ppm, and 60 ppm of phosphate as $PO_4$ (The Taylor high-phosphate slide comparator may be used instead.)

Four combination comparator mixing tubes, each marked 5, 15, and 17.5 ml, with stoppers

One filter funnel, 65-mm diameter

One package filter paper, 11-cm diameter

One 20-ml bottle

One 1/2-ml dropper

One 1/4-teaspoon measuring spoon

Two plain test tubes, 22 mm × 175 mm (about 50 ml)

Two rubber stoppers, no. 3 flask

One 250-ml glass-stoppered bottle, labeled *Comparator Molybdate Reagent*

The reagents you will need are

One 32-oz comparator molybdate

One 2-oz concentrated stannous chloride

One 32-oz standard phosphate test solution (45 ppm of phosphate, $PO_4$)

One 1-lb decolorizing carbon (This is a special grade of decolorizing carbon that has been tested to make sure it does not affect the phosphate concentration in the sample.)

For test purposes, the stannous chloride is supplied in concentrated form. The reagent must be diluted and should be prepared from the concentrated stannous chloride on the day it is to be used, because the diluted solution deteriorates too rapidly for supply by a central laboratory. If it is not fresh, dilute stannous chloride gives low test results. Concentrated stannous chloride also deteriorates over time, and should not be used if it is more than two months old. Dilute stannous chloride is made by the following method:

1. Fill the 1/2-ml dropper up to the mark with the concentrated stannous chloride.
2. Transfer it to a clean 20-ml bottle.

3. Add distilled water up to the shoulder of the bottle, then stopper and mix by shaking.

Any dilute stannous chloride not used the day it is made should be discarded.

To make the test for phosphate, here is the procedure to follow:

1. Without disturbing any settled sludge, transfer a sufficient amount of the sample to the test tube to fill it about half-full.

2. Add 1/4 teaspoon of decolorizing carbon. Stopper the tube and shake vigorously for about one minute. The carbon absorbs the tannin so that it can be filtered out.

3. Fold a filter paper and place it in the filter funnel. Do not wet down the filter paper with water. Filter the shaken sample, using a combination mixing tube as a receiver. The carbon absorbs tannin, and the tannin sludge present is filtered out more readily. Avoid jiggling the funnel, as unfiltered boiler water may flow over the edge of the filter paper and go into the test tube. It may be necessary to provide a support for the funnel. Filtering is slow because of the action of the carbon.

4. After 5 ml of the sample has been filtered through, as indicated by the level in the tube, discard it. Continue filtering to bring the level in the test tube up again to the 5-ml mark. The sample should come through clear and free, or nearly free, of any color resulting from the tannin. If not nearly free of tannin color, repeat the test using 1/2 teaspoon of carbon, added in two 1/4-teaspoon portions. Shake it for a minute after each addition.

5. Add comparator molybdate reagent to bring the level up to the second mark (15 ml). Stopper the mixing tube and mix by inverting the tube several times.

6. Add fresh dilute stannous chloride up to the third mark (17.5 ml). Stopper the mixing tube and mix by inverting. If phosphate is present, the solution in the mixing tube turns blue.

7. Place the tube in the comparator block. Compare the color of the solution in the tube with the standard colors of the phosphate color block. Colors between the two standard colors may be estimated. Take the reading within one minute after adding the stannous chloride, because the color fades quickly.

8. Record the results as *low,* if below 30 ppm; *high* if above 60 ppm; or *ok,* if between 30 and 60 ppm.

## Tannin Test

The purpose of the tannin test is to determine the amount of tannin in the boiler water. Tannin is used to hold sludge in suspension. In treating boiler water with tannin, the dosage can be controlled by controlling the depth of the brown color formed in the boiler water by the tannin. To estimate the depth of the color, which is necessary in adjusting tannin dosages, a sample of the boiler water is compared with a series of brown color standards of successfully increased depths of color. A tannin color comparator, which is used for the comparison, has five, glass color standards: No. 1 (very light), No. 2 (light), No. 3 (medium), No. 4 (dark), and No. 5 (very dark).

The kit for making the tannin test contains:

One tannin color comparator

Two square tubes, 13 mm viewing depth

One plain test tube, 22 mm × 175 mm

One filter funnel, 65 mm × 65 mm

One package of filter paper, 11-cm diameter.

In making this test, first fill a plain test tube almost to the top with cool boiler water. Then place a square test tube in the slot of the comparator, and insert the filter funnel in it. Fold a filter paper and place it in the funnel without wetting it down. Filter water from the plain test tube into the square tube until the tube is nearly half-full.

Remove the square tube from the comparator and hold it up to a good source of natural light. Note the appearance of the filtered boiler water. It should be free of suspended solids and sludge. If it is not, refilter the sample, using the same funnel and filter paper. Repeat, using double filter paper if necessary, until the sample does come through free of suspended solids and sludge.

To complete the test, place the square test tube of the filtered sample in the middle slot of the comparator. Then compare the color of the sample with the five standards, viewing against a good source of natural light. The color standard most closely matching the color of the filtered sample gives the tannin concentration in the boiler water. For a number of boiler water conditions, the tannin dosage is usually satisfactory if it maintains a medium (No. 3) tannin color. If the tannin color is too high, blowdown; if it is too low, add tannin.

## Test for Caustic Alkalinity (OH) without Tannin

The boiler water sample for this test is collected at a temperature of 70°F, or below.

The equipment required is as follows:

Two 8-in droppers with bulbs

Two 50-ml glass-stoppered bottles, labeled *Causticity No. 1* and *Causticity No. 2*

Four marked test tubes, 22 mm × 185 mm

Three plain test tubes, 22 mm × 175 mm

Three rubber stoppers, No. 2

One 14-inch test tube brush

One test tube clamp

Two 9-inch stirring rods

One 1-oz indicator dropping bottle for phenolphthalein indicator

One test tube rack

The following reagents are also required:

One 24-oz bottle causticity reagent No. 1

One 24-oz bottle causticity reagent No. 2

One 4-oz bottle of phenolphthalein indicator

Here are the steps to follow in conducting a test for causticity when tannin *is not* used. But first, as a word of caution: *Avoid exposure of the sample to the air as much as possible to minimize absorption of $CO_2$.*

1. Without disturbing any settled sludge, fill a marked test tube exactly to the first mark (25 ml) with some of the original boiler water sample.

2. Shake causticity No. 1 (barium chloride solution saturated with phenolphthalein) thoroughly and add enough to the graduated tube to bring the level exactly to the second, or long, mark (30 ml). Stir the solution with the 9-inch stirring rod, which must be kept clean and reserved for the causticity test only. If the mixture remains colorless or does not turn pink, the causticity in the boiler is zero. In this case the test is finished.

3. If the mixture turns pink, causticity is present. (If the pink color is not very deep, intensify it by adding two drops of phenolphthalein indicator to the mixture in the tube.) Add causticity reagent No. 2 (standard 1/30 normal acid), using the 8-inch dropper, which must be kept clean and reserved for the causticity test only. Causticity reagent No. 2 is sucked from the reagent bottle into the dropper by its rubber bulb and added, drop by drop, to the test tube. After each addition, stir the mixture with a stirring rod. After sufficient reagent has been added, the pink color just disappears, the change point usually being very sharp. As soon as the pink color just fades out, stop adding reagent.

4. The amount of causticity reagent No. 2 required to make the pink color disappear indicates the concentration of hydroxide (OH) or causticity in the boiler water. The amount of reagent used is shown by the marks on the test tube above the long mark (30 ml). The distance between any two marks on the test tube equals 5 ml, and readings less than 5 ml can be estimated. For example, if only 3/5 the distance between the long mark and the next mark above was filled, then 3 ml were added. If the distance filled was past one mark plus 3/5 the distance to the next, then $5 + 3 = 8$ ml were used. To obtain the actual ppm of hydroxide or causticity shown by the test, multiply the number of ml by 23. This constant number, 23, represents the amount of sodium hydroxide present in the boiler water by volume. Thus, for 8 ml of causticity reagent No. 2, there were $8 \times 23 = 184$ ppm hydroxide or causticity in the water.

5. Record the results of the test in a boiler log or a chemical log and adjust the range to meet the requirements. If the causticity is too high, blowdown; if it is too low, add sodium hydroxide (caustic soda).

## Test for Caustic Alkalinity (OH) with Tannin

For this test, it is desirable to start with a warm sample of about 160°F. It may be reheated by placing the sample-collecting container in a stream of hot boiler water drawn through the boiler water collector connection.

In a test for causticity when tannin is used, make sure you observe the following precaution as carefully as when tannin is not used: *Avoid exposure of the sample to the air as much as possible to minimize absorption of $CO_2$.*

The equipment and reagents required for this test are the same as those listed in the preceding section, where tannin was not used.

The procedure for conducting a test for causticity, when tannin is used, can be broken down into the following steps:

1. Fill two test tubes to the first mark (25 ml) with some of the original boiler water sample, taking care not to disturb the settled sludge in the container. (It is important that as little sludge as possible be transferred from the sample-collecting container to the test tubes.)

2. Shake the causticity reagent No. 1 thoroughly and add enough to each of the two marked test tubes to bring the levels up to the second, or long, mark (30 ml). Stir both with the stirring rod, which must be kept clean and reserved for the causticity test only. Stopper both tubes and let them stand until any sludge formed has settled to the bottom. The sludge carries down with it much of the tannin or other colored matter in the solution; settling takes a few minutes if the sample is warm. Without disturbing the sludge at the bottom, pour enough solution from the test tubes into the third marked test tube to fill it to the second, or long, mark. Discard the mixture left in the first two test tubes. If the sample in the third test tube is still warm, cool it by letting cold water run on the outside of the test tube. It is sometimes possible to intensify the pink color by adding two drops of phenolphthalein from the indicator dropping bottle to the sample in the test tube. Stir the solution. If it is not pink, the causticity in the boiler water is zero.

3. In that case, the test is finished. But if the mixture turns pink, proceed in the same manner as directed in steps (3), (4), and (5) when no tannin is used.

Perhaps a brief explanation should be given of an alternate procedure that can be followed in making the test for causticity when tannin is used. In this procedure, any glass container such as a large test tube or graduated cylinder, marked for 50 and 60 ml, can be used instead of two standard-marked test tubes used in steps (1) and (2) above. With the large test tube or graduated cylinder, the warm (160°F) sample is added to fill the cylinder up to the 50-ml mark and causticity reagent No. 1 is added to fill it up to the 60-ml mark. Stir the mixture and stopper the tube or graduated cylinder. After the sludge settles, pour off enough of the solution into one of the standard-marked test tubes to fill it to the long mark (30 ml). If the sample is warm, cool it by letting cold water run on the outside of the test tube. The pink color may be intensified by adding two drops of phenolphthalein. If the solution is not pink, the causticity in the boiler water is zero. But if it turns pink, proceed in the same manner as steps (3), (4), and (5) when no tannin is used.

## Test for Sodium Sulfite

The sample for this test should be cooled to 70°F, or below, and exposed to the air as little as possible, because oxygen in the air combines with sodium sulfite in the sample and causes low readings. It is desirable, therefore, to collect a separate sample, using the boiler water sample cooler, with the line reaching to the bottom of the sampling bottle. The boiler water should be allowed to run until a few bottlefuls overflow to waste.

The following equipment is needed for the sodium sulfite test:

Two marked test tubes
Two plain test tubes
One stopper for plain test tube
One stirring rod
One 8-inch dropper
One 1/4-teaspoon measuring spoon
One 50-ml beaker
One 150-ml beaker
One 30-ml acid dropping bottle, with dropper marked at 1/2 ml for hydrochloric acid 3N
One 30-ml starch dropping bottle, with dropper marked at 1/2 ml for starch indicator

The reagents required are as follows:

One 2-oz bottle of potato, or arrowroot starch
One 8-ml vial of thymol
One 24-oz bottle of hydrochloric acid 3N
One 1-pint amber bottle of standard potassium iodate-iodide reagent

The starch indicator for this test must be prepared locally. Here is the procedure to follow for good results.

1. Measure out a level 1/4 teaspoonful of potato or arrowroot starch and transfer it to the 50-ml beaker.

2. Add a few milliliters of distilled water and stir the starch into a thin paste, using the end of the stirring rod.

3. Put 50 ml of distilled water into the 150-ml beaker. (It is convenient in this step to have the 150-ml beaker marked at the point where it holds 50 ml, or one of the marked test tubes can be used by filling it with distilled water to the fourth mark above the long mark.)

4. Bring the water in the 150-ml beaker to a boil by any convenient method.

5. Remove the source of heat and immediately pour the starch paste into the boiling water while stirring the solution.

6. Put a crystal of thymol into the starch solution and stir. After the solution has cooled, pour off any scum on the surface and transfer 30 ml to the indicator dropping bottle.

The starch solution loses its sensitivity as an indicator after a time. Addition of the thymol preserves it for about two weeks. The starch should be dated when it is prepared.

In making the sodium sulfite test, proceed as follows:

1. Transfer 1 ml of hydrochloric acid 3N to a clean, marked test tube by measuring out 1/2-ml portions with the dropper of the acid dropping bottle.

2. From the starch dropping bottle, transfer 1/2 ml of starch to the marked test tube.

3. Without disturbing any settled sludge in the sample, pour enough sample into the marked test tube to bring the level up to the first mark (25 ml). Stir the mixture in the tube with the plunger end of the stirring rod.

4. To add the standard potassium iodate-iodide reagent to the mixture in the marked test tube, it is convenient to have the marked test tube supported and the stirring rod placed in the tube, so that the reagent can be added with one hand while the mixture is stirred with the other. Fill the 8-inch dropper with standard potassium iodate-iodide reagent from the stock bottle by sucking it up with the rubber bulb. (The dropper must be kept clean and reserved for this test only.)

5. Add the reagent to the mixture in the marked test tube, one drop at a

time, counting the number of drops and stirring after each is added until a permanent blue color, which is not removed by stirring, is obtained. The standard iodate-iodide reagent reacts with sodium sulfite in the mixture. The formation of the permanent blue color, because of the action of excess reagent with the starch, shows that all the sodium sulfite in the mixture has been consumed by the iodate-iodide reagent.

6. Each drop of iodate-iodide reagent used (except for the last one) indicates 5 ppm of sodium sulfite in the boiler water sample. To figure the concentration of sodium sulfite in the boiler water, multiply the total number of drops of the standard iodate-iodide reagent used, less one, by 5. For example, if 5 drops were used, subtract 1 from 5 = 4, 5 × 4 = 20 ppm.

7. Record the results of the test as ppm.

## Test for pH

The value for pH indicates the degree of acidity or alkalinity of a sample. A pH of 7.0 represents the neutral point; lesser values denote acidity, the greater values indicate alkalinity. The test is made as soon as possible after taking the sample. Avoid exposure to air as much as possible to minimize absorption of $CO_2$.

The equipment used in making the pH test of boiler water includes:

Two vials of indicator paper, hydrions pH 10 to 20
Two vials of indicator paper, hydrions C pH 11 to 12
One 50-ml beaker
One 2-oz bottle

In conducting the test for pH of boiler water, collect a 50-ml sample of the boiler water in the beaker. Remove a strip of pH 10-to-12 indicator paper from the vial and dip it into the sample in the beaker. Keep the paper immersed for 30 seconds, then remove it. If the sample does not change the color of the paper, or colors it yellow or very light orange, the pH of the sample is too low and the test is finished. If the paper turns orange or red, the pH is either satisfactory or too high.

In that case, remove a strip of paper pH 11-to-12 from the vial and dip it into the sample in the beaker. Keep the paper immersed for 30 seconds, then remove it. If the sample does not change the color of the paper, or colors it a light blue, the pH is satisfactory. If the paper turns deep blue, the pH is higher than necessary. In that case blowdown the boiler or reduce the dosage of caustic soda (NaOH).

## pH of Treated Condensate

In making a test for pH of treated condensate, take the sample from a point in the return piping near which condensation takes place, such as after a trap, or where return-line corrosion is known to occur. The sample must be representative of water flowing in the return lines. Water taken from the return tank, especially in large installations, generally shows a higher pH. A sample should not be taken from a collecting tank if other water, such as make-up water, is received in the tank.

The equipment required for this test includes:

One 4-oz bottle of condensate pH indicator
One 1-oz indicator bottle, with dropper marked at 0.5 ml
One 100-ml beaker, marked at 50 ml
One 9-inch glass stirring rod

In making a test for the pH of treated condensate, proceed as follows:

1. Pour a freshly drawn sample into the testing beaker until it is filled to the 50-ml mark. Cooling the sample is not necessary.
2. Transfer 1/2 ml of indicator solution to the 50-ml testing beaker, using the marked dropper. Stir the solution in the beaker.
3. If the color change is light pink, the sample is *neutral,* or slightly alkaline. The condensate pH is satisfactory and the test is over. Record in a log that the pH range is between 7 and 7.5.
4. If the color change is to green, the sample is in the acid range and the boiler water must be treated with amines, in small amounts at a time, and retested after each treatment. (Note: In some instances it may be necessary to get permission to treat the system with amines. Amines are volatile, poisonous, and in the alkaline range. Amines are the only chemical used to treat boiler water that will vaporize and leave with the steam and thereby protect the return system.)

If the color should change to either red or purple, the sample is in an excessive alkaline (pH) range. In that case, reduce the amines treatment in small amounts at a time, and retest after each treatment. Remember, the normal acceptable condensate pH range is between 7 and 7.5.

## Test for Total Dissolved Solids (TDS)

The solu-bridge method is a simple and rapid way to determine TDS content. It is based on the fact that ionizable solids in water make the solution conduct electricity. The higher the concentration of ionizable salts, the greater the electrical conductance of the sample. Pure water, free from ionizable solids, has a very low conductance and thus a very high resistance. The solu-bridge instrument measures the total ionic concentration of a water sample, the value of which is then converted to parts per million.

The solu-bridge test equipment and reagent are furnished by the supplier of the kit.

*Caution:* The model of the solu-bridge given below is not suitable for measuring solids in condensed steam samples or an effluent of the demineralizing process. Instead, a low-conductivity meter is necessary because of the extremely low solids content of good condensed steam and demineralized water.

The equipment and reagent for making the test are:

One solu-bridge, Model RD-P4 or equivalent, for a 105- to 120-v, 50- to 60-hertz a-c electrical outlet (The model has a range of 500 to 7,000 micromhos/cm.)

One polystyrene dip cell, Model CEL-S2

One thermometer, 0° to 200°F

One 0.1-g dipper for gallic acid

One cylinder, marked at 50-ml level

Gallic acid powder, 1 lb

Calibration test solution, 1 qt

The test is made as follows:

1. Without shaking, pour 50 ml of the sample into the cylinder. Add two dippers of gallic acid powder and mix thoroughly with a stirring rod.
2. Connect the dip-cell leads to the terminals of the solu-bridge and plug the line cord into a 120-v a-c electrical outlet. Turn the switch to *on,* and allow the instrument to warm up for one minute.

3. Clean the cell by moving it up and down several times in distilled water. Measure the temperature of the sample to be tested, then set the pointer of the solu-bridge temperature dial to correspond to the thermometer reading.

4. Place the cell in the cylinder containing the 50-ml sample. Move the cell up and down several times under the surface to remove air bubbles inside the cell shield. Immerse the cell until the air vents on the cell shield are submerged.

5. Turn the pointer of the solu-bridge upper dial until the dark segment of the tube reaches its widest opening.

6. Calculate the result in ppm by multiplying the dial reading by either 0.9 or by a factor recommended by the local instructions. For example, if the dial reading is 4,000 micromhos and the factor used is 0.9, then $4,000 \times 0.9 = 3,600$ ppm.

7. Record the results of the test as ppm.

# FEED WATER SYSTEM

A typical feed water system may include from one to several components which are piped together to permit the introduction of treated water into a boiler water system.

A reliable feed water treatment consultant should be contracted to analyze the feedwater and make recommendations for the proper treatment of the water to prevent:

1. Corrosion
2. Scale-forming deposits
3. Carryover

## Feed Water System Components

Some of the major components which could be included in a feed water system are (1) water softeners, (2) dealkalizers, (3) deaerators, (4) surge tanks, (5) feed water systems, (6) chemical feed systems, manual shot feeders, (7) reverse osmosis, (8) commercial reverse osmosis, (9) continuous blowdown, heat recovery, (10) flash tank heat exchangers, (11) blowdown steam-water separators, (12) filters, and (13) sample coolers.

**Water Softeners.** Water softeners are required where the water is naturally hard. (See Figure 10–1.) The calcium and magnesium compounds in water are the factors which cause it to be hard. These compounds are relatively insoluble in water and have a tendency to stick to the metal surfaces of the boiler system in the form of scale and other deposits. Water softeners remove the hardness elements and replace them with highly soluble sodium ions. This prevents the buildup of scale on the heat transfer surfaces and helps to maintain peak boiler efficiency. Removing this hardness also reduces the need for chemical treatment used to control scale. The factors that determine the size of the water softening equipment are

1. Maximum water flow rates
2. Amount of make-up water needed

**FIGURE 10-1.**
Water softener equipment (Courtesy of Cleaver-Brooks)

3. Hours of operation

4. Water hardness level

**Dealkalizers.**   The process used for reducing the alkalinity of water is called dealkalization. Dealkalizers are used to reduce fuel costs due to frequent blowdowns where alkalinity is a problem. Dealkalizers require that softened water be used. They are, therefore, installed downstream from the water softener and before the deaerator. (See Figure 10–1.)

Chloride cycle dealkalizers are similar to water softeners except that, instead of sodium zeolite resin, they use a strong anion resin to remove negatively charged ions from the raw water supply. These include bicarbonate, carbonate, sulfate, nitrate, and silica. Those removed ions are replaced with chloride ions. (See Figure 10–2.)

Dealkalizers remove more than 90% of the bicarbonate alkalinity from the softened water supply. This effectively controls the formation of carbon dioxide in the boiler water system. Carbon dioxide is the major cause of condensate line corrosion. Dealkalization of the feed water reduces the need for neutralizing chemicals by as much as 90%, and cuts fuel costs by minimizing the amount of blowdown required.

**FIGURE 10-2.**
Typical dealkalizer (Courtesy of Cleaver-Brooks)

**Deaerators.**    Deaeration is widely recognized as the most acceptable method for the removal of oxygen and carbon dioxide from the boiler make-up water and condensate. Therefore, a deaerator should be seriously considered with every boiler installation. (See Figure 10–3.) Without deaeration, these dissolved gases can cause serious corrosion problems. This corrosion appears as erosion and pitting, and eventually will result in total perforation of the metal surfaces. The effects are costly in terms of poor fuel efficiency, frequency of maintenance, outages, and premature boiler failure.

Deaerators are considered essential for:

1. All boiler plants operating at 75 psig or more
2. All boiler plants with little or no standby capacity
3. All boiler plants where production depends on continuous operation
4. All boiler plants operating with cold water makeup of 25% or more

In addition to removing oxygen and carbon dioxide, deaerators heat the boiler feed water, which helps to reduce boiler maintenance caused by thermal shock. They may also provide a means for recovering heat from exhaust steam and hot condensate.

**FIGURE 10-3.**
Typical deaerator (Courtesy of Cleaver-Brooks)

There are basically three types of deaerators.

1. Spray deaerators
2. Tray deaerators
3. Packed column deaerators

**Surge Tanks.**   Surge tanks can greatly reduce the dependence on cold, untreated raw water to replace boiler system losses by collecting condensate by gravity return for re-use in the boiler feed system. Saving water will reduce the treatment costs by recycling condensate that has already been treated. (See Figure 10–4.)

Surge tanks are required when intermittent peak loads of condensate can exceed the storage capacity of the deaerator.

Savings result from the continous flow of hot water condensate into the surge tank. Integrated control automatically introduces cold water makeup to supplement the condensate only when necessary to meet boiler demand. This process reduces the amount of heat required to heat the boiler feed water.

**Boiler Feed Water Systems.**   Boiler feed water systems help to maintain peak efficiency and prolong the life of boilers when investment in a deaerator cannot be justified. (See Figure 10–5.) Consisting of one or more feed pumps and a corrosion-resistant receiver tank, the systems automatically supplement the condensate with make-up water to replace system losses. The cold water is heated by mixing with the hot condensate and is then pumped to the boiler on demand.

With an automatic preheater, the feed water temperature can be maintained at 210°F. At this higher temperature, substantial oxygen and carbon dioxide are released from the water. This protects the boiler from corrosion problems. Preheating is recommended if the return condensate constitutes 50% or less of the feed water required.

Another advantage of high temperature feed water is that it helps eliminate any problems caused by thermal shock. It does this by reducing expansion and contraction of the boiler components to extend boiler life. It also makes it easier to maintain boiler pressure during peak loads or load swings.

**Chemical Feed Systems, Manual Shot Feeders.**   Chemical feed systems treat feed water and condition the blowdown sludge when an extra amount of protection against corrosion and scaling is required. Automatic systems

**FIGURE 10-4.**
Typical surge tank (Courtesy of Cleaver-Brooks)

are available for single- or multiple-boiler installations. (See Figure 10–6.) They are fully packaged and can be equipped with either piston or diaphragm-type pumps capable of accurate chemical injection from 0 to 6.2 gal/hr at a pressure of up to 1,000 psig.

Manual shot feeders are used for batch feeding of chemicals into closed

**FIGURE 10-5.**
Typical boiler feed water system (Courtesy of Cleaver-Brooks)

loop or low make-up water systems. They are completely assembled and ready for installation. (See Figure 10–7.)

**Reverse Osmosis.**    Reverse osmosis is a continuous high-pressure membrane separation process that has been proven to deliver high-purity boiler feed water. (See Figure 10–8.) These units remove up to 95% of the total dissolved solids from the feed water. The only electrical power used is to power a repressurization pump.

Reverse osmosis systems typically reduce the chemical treatment costs from 50% to as much as 75%. They can also eliminate hazardous waste

**FIGURE 10-6.**
Automatic chemical feed system
(Courtesy of Cleaver-Brooks)

problems associated with processes requiring large amounts of chemicals to achieve the same high purity standards.

These systems are particularly suited to situations where make-up water is high in dissolved solids, alkalinity, and silica. Support equipment requirements are very small, normally consisting of a water softener and a filter. They are generally self-regulating, and can be operated automatically.

**FIGURE 10-7.**
Manual shot feeder (Courtesy of Cleaver-Brooks)

**FIGURE 10-8.**
Reverse osmosis unit
(Courtesy of Cleaver-Brooks)

**Continuous Blowdown, Heat Recovery.**  A continuous boiler surface blowdown is the most effective method for purging destructive solids from any steam system. (See Figure 10–9.) The heat recovery feature is a benefit because without it only system protection is achieved. The heat loss still occurs. The integrated system automatically adjusts to changing demand and recovers 90% of the heat normally lost through surface blowdown. The system operates automatically, with the heat recovery units regulating the blowdown flow to accommodate changing boiler makeup loads. A modulating flow controller prevents overheating of makeup to avoid scaling and damage to the heat exchanger.

**Flash Tank Heat Exchangers.**  Flash tank heat exchangers are used when the operating pressures exceed 250 psig and a continuous blowdown recovery is desired. They consist of a flash tank and a blowdown heat exchanger. (See Figure 10–10.) Water from the continuous blowdown enters the tank and separates into steam and water. The steam, normally at 5 psig, can be returned to the deaerator to help meet the heating needs. Hot water leaves the flash tank at temperatures of 220°F or more. It goes into the heat exchanger, where it heats cold make-up water entering the system.

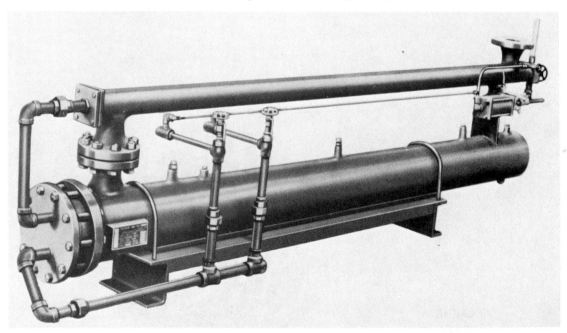

**FIGURE 10-9.**
Continuous blowdown, heat recovery unit (Courtesy of Cleaver-Books)

**FIGURE 10-10.**
Flash tank heat exchanger (Courtesy of Cleaver-Brooks)

**Blowdown Separators.**   The boilers that supply steam for power, process, or heating applications require periodic, and more often frequent, blowdowns to prevent buildup of harmful solids. This protects the boiler from severe scaling or corrosion problems that would otherwise result.

Blowdown separators use a safe, economical flash purification process for enhancing blowdown effectiveness. (See Figure 10–11.) The steam is rapidly separated from the blowdown water and is vented out the top in a cyclonic spinning action. The water and dissolved solids are flushed out the bottom drain. The internal pressures do not exceed 5 psig. The blowdown water is cooled to 120°F by a drain tempering device designed to meet state and local codes.

**Filters.**   The need for special filtering equipment for the removal of impurities can be readily determined by raw water analysis. Where filters are recommended, the cost is quickly recovered by eliminating the need for frequent equipment cleaning and servicing, inefficient performance, damage to the system components, and premature equipment failure.

Most filters are capable of removing particles down to 10-micron size.

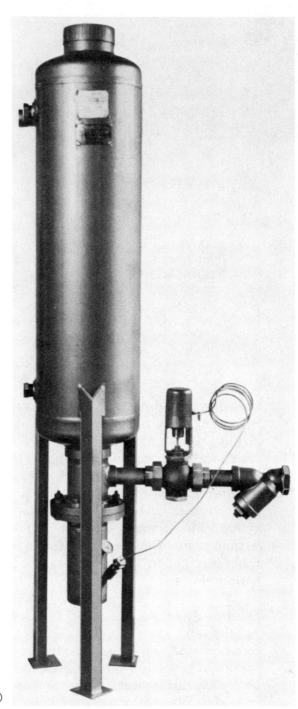

**FIGURE 10-11.**
Blowdown separator
(Courtesy of Cleaver-Brooks)

They remove sediment, suspended solids, and colloidal matter down to the 10-micron size from the boiler make-up water. (See Figure 10–12.) They are highly effective in controlling sludge buildup. Periodic backwashing is required to remove the collected matter from the filter bed.

Sand filters can handle heavy loads in controlling sludge buildup. They remove heavy, large-particle settlement and suspended solids down to the 40-micron size from the make-up water. Periodic backwashing is required to remove the collected matter from the filter bed.

Activated carbon filters remove free chlorine, some dissolved organics, and sediment down to the 40-micron particle size from the boiler make-up water. By removing sediment, activated carbon filters also help control sludge buildup in boilers. They protect pretreatment equipment by

1. Removing free chlorine (which is corrosive and attacks the ion-exchange resin cross-linking agent in water softeners and dealkalizers, and the membranes in reverse osmosis equipment)
2. Removing organics (which promote fouling in ion-exchange beds, can contaminate water supplies, and cause foaming in boilers and heat exchange equipment)

Periodic backwashing is required to remove collected matter from the filter bed.

Iron filters are designed to remove both soluble and insoluble iron, suspended solids and colloidal matter down to the 40-micron particle size

**FIGURE 10-12.**
Filter (Courtesy of Cleaver-Brooks)

from boiler make-up water. They are used to control sludge buildup caused by iron precipitation. The water pH must be at least 6.2. Occasional regeneration with potassium permanganate may be required.

**Sample Cooler.**    Sample coolers provide a low-pressure drop system for monitoring boiler water quality without shutting down. (See Figure 10–13.) The sample cooler is simple, compact, and designed to cool boiler samples with minimum disturbance to the system pressure.

The design has a baffling on the shell side to achieve high-cooling water flow, which minimizes pressure drop, vibration, and scaling. The shell can be easily removed for inspection and/or cleaning without disconnecting sample or water lines.

**FIGURE 10-13.**
Cooling chamber
(Courtesy of Cleaver-Brooks)

# BOILER WATER TESTING

(This section is reprinted by permission of the American Boiler Manufacturers Association). Monitoring the various water systems is an important part in the successful operation of steam or hot water boiler systems. Failure to do so may cause unscheduled equipment failure, or lead to operating problems which could otherwise be avoided. The testing may be either chemical or mechanical, or a combination of both. The chemical tests most commonly employed are as follows:

| *Test* | *Equivalent Reported* |
| --- | --- |
| Phenolphthalein Alkalinity | $CaCO_3$ |
| Methyl Orange Alkalinity | $CaCO_3$ |
| Chloride | Cl or NaCl |
| Total Hardness | $CaCO_3$ |
| pH | Expressed in units |
| Soluble Phosphate | $PO_4$ |
| Sulfite | $SO_3$ or $Na_2SO_3$ |
| Total Dissolved Solids | Micromhos or ppm |

Modern chemical analysis will usually be reported as parts per million. In some cases, the older term grains per U.S. gallon is employed. One gr/gallon equals 17.1 ppm.

A measure of the hydrogen ion concentration, pH covers a range of 0-14. A pH of 7.0 is neutral. Above pH 7, the solution is alkaline, below pH 7, it is acidic.

## Steam Boiler Systems

The different water circuits in a steam boiler system include (1) make-up water, (2) feed water, (3) boiler water, and (4) condensate. The following is a discussion of the testing for these different circuits.

**Make-Up Water.**   The total hardness test is used to monitor the operation of the water softener. The effluent hardness will begin to rise as the end of the useful softening cycle is reached. Determination of the hardness establishes when regeneration should be initiated. The Versene hardness test has replaced the older soap method because it is much simpler, more accurate and less time consuming.

The alkalinity test is used to monitor chlorine anion dealkalization. In this case, the alkalinity of the effluent increases as the ion exchange capacity reaches its limit. Regeneration of the dealkalizer is initiated when the alkalinity reaches some predetermined point. Alkalinity is also normally used to determine when the unit has been rinsed sufficiently and is ready to be returned to service.

The chloride is measured after regenerating a sodium water softener. The purpose is to make sure that the unit has been rinsed sufficiently and is ready to be returned to service. High chloride concentrations indicate that the brine has not been completely rinsed. Alternate tests which may be used in controlling the rinse cycle are total dissolved solids and total hardness.

**Feed Water.**    The total hardness determination serves many purposes. If the raw water is not softened, it can be used to determine the percentage of raw water in the feed water. Where a softener is used to remove hardness, it serves to indicate contamination caused by unsoftened water.

Chloride or total dissolved solids are also used to check the feed water quality. They would normally be employed to detect contamination from a source other than the raw water.

**Boiler Water.**    The phenolphthalein alkalinity is used to control the feed of caustic soda or the rate of blowdown. Problems with foaming or priming may be attributed to high alkalinity. While not necessarily true, it is usually prudent to avoid excessive concentrations.

Chloride or conductivity are normally used to control the boiler water solids concentration through blowdown. Like alkalinity, abnormally high solids concentrations may promote the tendency for foaming or priming.

Sodium sulfite is the most commonly used chemical oxygen scavenger in systems operating below 400 psig (2859.35 kPa). Determination of the sulfite residual in the boiler water is used to control the feed of this chemical.

The importance of maintaining proper control over the scale-inhibiting chemicals cannot be overemphasized. The water treatment consultant will recommend the tests which are to be used in maintaining proper chemical treatment concentrations.

**Condensate.**    The pH measurement is used to control the feed of neutralizing amines which are commonly employed to control corrosion. A pH meter or the use of pH indicators is acceptable.

Condensate conductivity will normally be low because it is relatively pure water. Increases in conductivity will reflect mineral concentration. However, additional analyses will usually be required to clearly define the source of contamination.

Corrosion monitoring is important when using filming amines for corrosion control. This is because there is no simple, suitable test which is employed in controlling the feed of filming amines. Corrosion studies are also desirable when applying neutralizing amines. Rectangular specimens are preferred because the nature of corrosion can be more clearly observed with this type of coupon. (See Figure 10–14.)

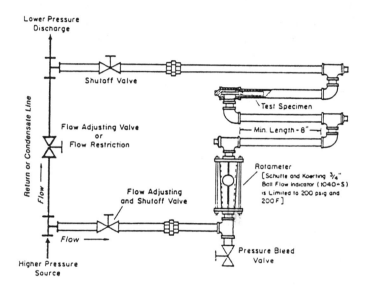

**NOTES**

1. All piping and valves for the bypass are to be ¾ inch iron or steel.
2. To minimize any tendency for the accumulation of sediment or the entrapment of air around test strips, install the bypass in vertical line, around two points where there is a pressure differential.
3. Close shut off valves and open bleed valve before removing any test specimens.
4. Test specimens must be handled carefully to avoid coating with natural oil from the skin, with thread compound or any similar material.
5. The end of the pipe plug holder is notched parallel to the flat side of the test specimen. When installing, the notch (consequently the specimen) is to be positioned vertically.
6. As a rule, all specimens are to be installed at the same time. Removal is to be at intervals to measure the effect of exposure time.
7. Adjust flow through the bypass to 3-5 gallons per minute, equivalent to 1.8-3.0 feet per second with ¾ inch pipe.
8. When removed, each specimen is to be carefully disconnected from the holder and then immediately dried with a blast of hot air or by sponging with a paper tower or tissue. Do not clean.
9. Analysis of corrosion tests can be obtained from your contracted water consultant.

**FIGURE 10-14.**

Recommended corrosion test specimen closed-loop bypass installation (Courtesy of The American Boiler Manufacturers Association)

## Hot Water Boiler Systems

The different circuits in a hot water boiler system include (1) make-up water, and (2) boiler water. The following is a discussion of these different circuits.

**Make-up Water.**    The quantity of make-up water used is usually more important than the quality (hardness). Scale formation is not ordinarily a problem in systems which do not have significant water losses. A suitable monitoring device which measures the quantity of make-up water used not only indicates when unknown increases in water loss occur, but may also prove helpful in establishing whether or not the quality of the make-up should be improved through the use of softening equipment.

**Boiler Water.**    The hardness of the boiler water will normally stabilize at some average level which will be related to the make-up water characteristics and operation of the system. An increase in the hardness would tend to indicate higher than normal make-up water usage and increase the potential for scale formation.

The pH should be monitored because it will be desirable to maintain alkaline conditions to minimize the potential for corrosion. (Note: These levels should be established by operation and the experience of the water treatment consultant.)

The concentration of boiler water treatment chemicals should be closely monitored. Such materials would usually be employed to control corrosion and/or scale formation.

Corrosion monitoring is also desirable. This serves to insure that the optimum conditions are being maintained in the boiler water.

## Condensate System Protection

The major causes of corrosion in the condensate system are (1) carbon dioxide, and (2) oxygen. The following is a discussion of the causes.

**Carbon Dioxide.**    The corrosion in steam/condensate return systems is usually caused by carbon dioxide. This is because it dissolves in the water formed when condensation occurs, producing carbonic acid. Corrosion caused by carbonic acid will occur below the water level, and it can be identified by the channeling or grooving of the metal wall.

Breakdown of the make-up alkalinity is the source of carbon dioxide.

This occurs as the water is heated in the boiler. If the amount of carbon dioxide in the steam is to be reduced, then treatment of the make-up water is required to reduce the alkalinity.

**Oxygen.**   Oxygen can also cause corrosion in condensate systems. However, it is not as common as carbon dioxide corrosion because oxygen can be removed from the feed water through temperature elevation in a hot well or a deaerating heater and by the use of chemical oxygen scavengers. Oxygen corrosion occurs wherever moisture is present, so it will be found both above and below the water level in the piping. It is easily recognized because of the characteristic pitting of the metal.

Severe oxygen corrosion usually occurs in vacuum return systems. Leaks permit air to be drawn into the system. Effective maintenance of the system is required to properly control this problem. Corrosion due to oxygen also occurs in systems which are operated intermittently. There usually is no simple mechanical solution to this problem.

## Treatment

The treatment of the condensate system involves the following chemicals: (1) neutralizing amines, (2) filming amines, and (3) organic oxygen scavengers. These chemicals are discussed below.

**Neutralizing Amines.**   These are organic compounds which are used to neutralize the acidity of the condensate. The corrosivity is greatly reduced by maintaining the condensate pH in an alkaline range. A pH somewhere between 8.0 and 9.0 is usually recommended. The severity of oxygen corrosion is also reduced at this high pH.

The properties of neutralizing amines vary with respect to their neutralizing ability and also the manner in which they are distributed throughout the steam distribution system. Accordingly, they may be used alone or in combination to achieve the desired results. Cyclohexylamine, diethylaminoethanol, and morpholine are the most commonly used neutralizing amines. This is primarily because they are approved for use in plants which are regulated by the Federal Drug Administration (FDA) and the U.S. Department of Agriculture (USDA).

**Filming Amines.**   Unlike neutralizing amines, filming amines do not neutralize acidity or raise pH. Rather, they form a nonwettable barrier which repels the corrosive environment of the condensate. They are known as polar amines because one end of the molecule attaches itself to the metal surface when condensation occurs. Filming amines can handle severe oxygen corrosion problems more effectively than neutralizing amines.

The properties of filming amines also vary, particularly with respect to the corrosion resistance of the water repellent film. Combinations of filming and neutralizing amines are often employed. Octadecylamine is the only filming amine approved for application in FDA/USDA regulated plants. Unlike neutralizing amines, it is preferable to inject filming amines directly into the steam header. This is largely because the boiling point of these materials usually exceeds that of the boiler water.

*Caution:* In existing systems, the gradual introduction of filming amines is recommended to avoid the rapid removal of corrosion products. Fast removal of rust will cause plugging of traps and strainers.

**Organic Oxygen Scavengers.**   The volatile or organic oxygen scavengers are employed as supplementary treatments, particularly where oxygen is causing a corrosion problem. Organic oxygen scavengers have the ability to establish and maintain a corrosion-resistant film on the metal surfaces. This is usually called metal conditioning. In condensate treatment, these materials are not usually fed to completely remove oxygen from the condensate.

## Monitoring

The sample point or method of sampling can influence the results. If the condensate water is hot or under pressure and flashes, then carbon dioxide may be lost. This raises the pH. Accordingly, it is preferable to cool the condensate sample to room temperature.

It is not absolutely necessary to use a completely representative test or sample to assure success in effectively regulating a condensate corrosion control problem. Rather, it is the consistency in sampling and testing that is important.

Corrosion monitoring is important when using filming amines for corrosion control. This is because there is no simple, suitable test which is employed in controlling the feed of filming amines. When using organic scavengers, such studies help determine if the metal surfaces are being effectively conditioned. Corrosion studies are also desirable when applying neutralizing amines. The bypass installation shown in Figure 10–14 is normally used for this purpose. Rectangular or oblong specimens with a flat surface are preferred because the nature of corrosion can be more clearly observed with this type of coupon.

# REVIEW QUESTIONS

1. Is water that is acceptable for domestic purposes necessarily good for use in boilers?
2. What do scale deposits consist of?
3. What factors determine the water treatment method used for boilers?
4. What causes corrosion in boilers and piping?
5. Under what conditions is external water treatment usually used?
6. What is the purpose of tannin when used in boiler water treatment?
7. What test is based on the determination of the total calcium and magnesium content of the boiler water?
8. When testing boiler water for hardness, what should the sample color change be?
9. What happens to the tannin when testing boiler water for phosphate?
10. How long should prepared stannous chloride be stored?
11. What color will the test solution be if phosphate is present?
12. In parts per million, when is the phosphate level of boiler water considered to be good?
13. How is the amount of tannin in boiler water determined?
14. When is the tannin content of boiler water satisfactory?
15. What precaution should be taken when testing boiler water for causticity?
16. When causticity is present, what color will the boiler water sample turn?
17. What should be the maximum temperature of the sample when testing for sodium sulfite?
18. What is the purpose of putting thymol in a starch solution?
19. At what pH is a water sample considered to be neutral?
20. When checking the pH of boiler condensate, what water should be represented by the sample?

# CHAPTER 11

# STEAM CONTROLS

## Learning Objectives

The objectives of this chapter are to:

- Describe different types of controls used in steam boiler installations
- Show the operation of the different types of steam controls used on steam boilers
- List proper installation practices used when installing steam boiler controls
- Make the reader aware of the purposes of the different safety-type controls used on steam boilers.

When automatic controls are used, boiler operation is safer and more economical than when manual controls are used. Automatic controls can maintain the proper amount of water in the boiler. They can regulate burner operation in response to the demands placed on the boiler and they can do this with little supervision.

There are many different types of automatic controls used on steam boilers. Some of them are single-purpose while others are multi-purpose controls.

It is the responsibility of the boiler operating engineer to know which types are used on the boiler he is overseeing. He must know how they operate and how to check them for proper operation.

## OPERATING PRESSURE CONTROL

This is a pressure-actuated control that is mounted on the steam outlet of the boiler. It measures the steam pressure at that point and signals the burner to either operate or to stop operating. (See Figure 11-1.)

Operating pressure controls are connected to the boiler with a siphon (pigtail) between them and the boiler, with the loop either towards the front or the rear of the control so that expansion and contraction of the metal will not affect their setting. (See Figure 11-2.) The exposed (uninsulated) coil is provided in the line leading to the control, and the steam condenses into water in this exposed coil of tubing. Thus there is always a condensate

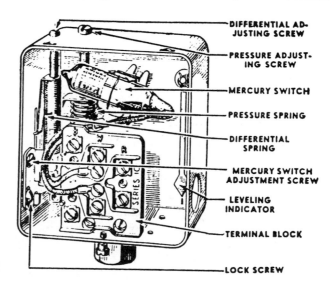

**FIGURE 11-1.**
Operating pressure control

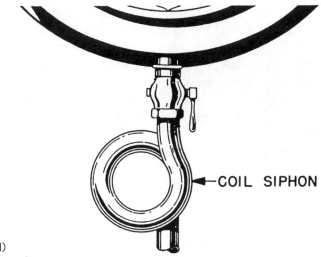

**FIGURE 11-2.**
Siphon (or pigtail)

seal between the control and the steam source. Sometimes the condensate seal is lost when repairs are made to the control or to the control line. In this event, the loop must always be filled with water before the control is subjected to steam line pressure.

In operation, when the steam pressure drops to a predetermined point the operating control signals the burner that steam is needed. The burner starts heating the boiler and when the steam pressure has risen to the desired pressure, the control signals the burner to stop operating. The pressure settings will be determined by the design and use of the boiler and its connecting system.

## HIGH-PRESSURE LIMIT CONTROL

The high-pressure limit control is the same type of control as the operating pressure control. It is usually mounted in approximately the same location and has the same mounting requirements. The purpose of the high-pressure control is to stop the burner in the event of high pressure inside the boiler. This prevents damage to the boiler, and the possibility of an explosion. The cutout setting of the control used for this purpose is several pounds higher than the operating pressure control. It is a safety control that seldom operates.

# GAUGE GLASSES

Each boiler must have at least one water gauge glass; some types require more than one depending on their operating pressure. (See Figure 11-3.) The gauge glass allows the operater to see the water level in the steam drum. Each gauge glass must have a valved drain, and the gauge glass and pipe connections must be no less than 1/2-inch pipe size. The lowest visible part of the water gauge must be at least 2 inches above the lowest permissible water level, which is defined as the lowest level at which there is no danger of overheating any part of the boiler when it is operating at that level. Horizontal fire-tube type boilers are set to allow at least 3 inches of water over the highest point of the tubes, flues, or crown sheet at the lowest reading in the gauge glass.

Each water gauge glass consists of a strong glass tube which is connected to the boiler or water column by two special fittings. These fittings sometimes have an automatic shutoff device, usually a nonferrous ball that functions if the water glass should fail. The automatic shutoffs must be made

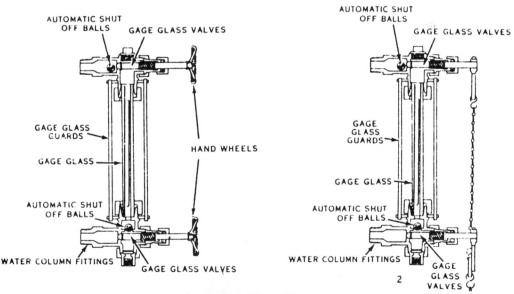

1.  Hand Shutoff
2.  Automatic Shutoff

**FIGURE 11-3.**
Cross sections of typical gauge glasses

in accordance with certain provisions of the boiler code. The following are some of these provisions:

1. The top ball must not be capable of completely shutting off the flow.
2. The bottom ball must rise vertically to the seat.
3. The balls must be accessible for inspection.
4. The bottom ball must be able to be removed and inspected while the boiler is in operation.

# TRY COCKS

Unless two gauge glasses are installed on the same horizontal line, each boiler must have three or more try cocks. (See Figure 11–4.) These try cocks must be located within visible length of the gauge glass. Boilers require only two try cocks if they are not over 36 inches in diameter and the heating surface does not exceed 100 square feet. Gauge cocks are used to check the accuracy of the gauge glass. They are opened by handwheel, chain wheel, or lever, and are closed by hand, by a weight, or spring.

# WATER COLUMN

A water column is a hollow vessel having two connections to the boiler. (See Figure 11–5.) The top connection enters the steam drum of the boiler through the top of the shell or drum. The water connection enters the shell or head at least 6 inches below the lowest permissible water level. The purpose of the water column is to steady the water level in the gauge glass through the

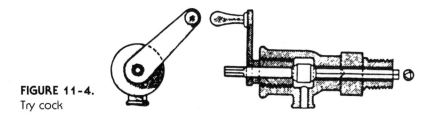

**FIGURE 11–4.**
Try cock

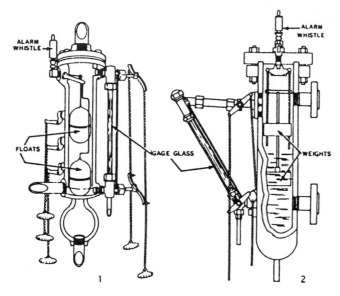

1.  With float type high - and low water alarms
2.  With weight type high - and low water alarms

**FIGURE 11-5.**
Cross sections of typical water columns

reservoir capacity of the column. Also, the column may eliminate the obstruction of small diameter, gauge glass connections by serving as a sediment chamber.

The water columns illustrated are equipped with high- and low-water alarms which sound a whistle to warn the operator. The whistle is operated by either the two floats or the solid weights shown in Figure 11-5.

## WATER LEVEL CONTROL

The water level control not only automatically operates the boiler feed pump; it also safeguards the boiler against low water by stopping the burner. Various types of water level controls are used on boilers. They may be equipped with a float-operated type, an electrode probe type, or a combination float and mercury switch type of automatic water level control. Each of these types is described below.

## Float-Operated Type

The float-operated type of feed water control may be attached to the water column or to special fittings on the boiler. (See Figure 11-6.) This control accomplishes its purpose by means of a float, arm, and a set of electrical contacts. As a low water cutoff, the float rises or lowers with the water level in the boiler. The control is connected by two lines which allow the water and steam to have the same level in the float chamber as in the boiler. An arm and linkage connect the float to a set of electrical contacts, which operates the feed water pump when the water lowers the float. If the water supply fails or the pump becomes inoperative and allows the water level to continue to drop, another set of electrical contacts operates an alarm bell, buzzer, or whistle, and stops the burner operation.

**FIGURE 11-6.**
Feeder cutoff combination (Courtesy of McDonnell & Miller, ITT Fluid Handling Division)

## Combination Float and Mercury Switch Type

This type of water level control reacts to changes made within a maintained water level by breaking or making a complete control circuit to the feed water pump. It is a simple two-position type control, having no modulation or differential adjustment or setting. As all water level controllers should be, it is wired independently from the programmer. The control is mounted at steaming water level and consists of a pressurized float, a pivoted rocker arm, and a cradle-attached mercury switch.

The combination float and mercury switch type of water level control functions as follows: As the water level within the boiler tends to drop, the float lowers. As the float lowers, the position of the mercury switch changes. Once the float drops to a predetermined point, the mercury within the tube runs to its opposite end. This end contains two wire leads, and when they are connected through the puddle of mercury, an electric circuit to the feed water pump is completed. The pump, being energized, pumps water into the boiler. As the water level within the boiler rises, the float rises. As the float rises, the position of the mercury switch changes. Once the float rises to a predetermined point, the mercury runs to the opposite end of its tube, breaking the circuit between the two wire leads and stopping the feed water pump. The feed water pump remains off until the water level again drops low enough to trip the mercury switch.

## Electrode Probe Type

The electrode probe type of feed water regulating and low water cutoff control consists of an electrode assembly and a water level relay. The electrode assembly contains three electrodes of different lengths corresponding to high, low, and cutoff water level in the boiler drum. (See Figure 11–7.)

The electrode assembly is installed on the top of the boiler drum so that the normal water level is at approximately the midpoint between the high-level and the low-level electrodes. The electrodes are electrically wired to the water level relay assembly. The relay contains the feed water pump start and stop contacts, which are wired to the feed water pump controller. The low-water cutoff contacts are wired to the boiler burner circuit. The feed pump is started when the water level in the boiler drops below the middle electrode. After the water level is restored and reaches the high-level electrode, the feed pump is stopped. In the event that the water level is not restored and drops below the longest electrode, the cutoff contacts of the

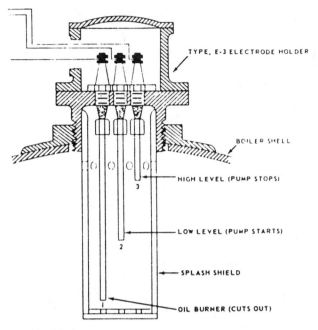

**FIGURE 11-7.**
Electrode-type water level control

water level relay stop the boiler burner by breaking the burner circuit. The electrodes are held in position by screwing them into a threaded connector, located at the top of the electrode bottle. In some cases they also are locked in position with a suitable locking device; however, this is not a design feature common to all installations. Boiler casualties have occurred from low water cutoff electrodes backing off from the threaded connectors, and grounding with the assembly bottle. On those installations where one is not present, a locking device should be installed.

## PRESSURE GAUGES

Each boiler is equipped with a pressure gauge, which serves as a means of measuring pressure inside the boiler. There are several types of pressure gauges in common use, including the Bourdon tube type. (See Figure 11–8.)

The Bourdon tube is made of seamless metal tubing and is oval in cross section. The tube is bent in the form of the letter C. One end of the tube is

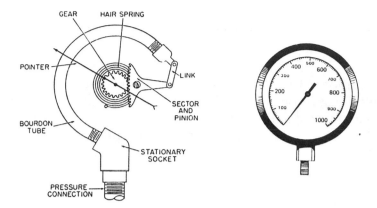

**FIGURE 11-8.**
Bourdon tube pressure gauge

open and is connected to a socket that opens into the pressure line leading to the gauge. The other end of the Bourdon tube is closed, and is free to move; this free end is connected to the gauge pointer by a linkage system of levers and gears. The position of the free end of the Bourdon tube represents a balance between the two forces—the pressure of the liquid or gas inside the Bourdon tube which tends to straighten the tube, and the spring action of the metal which tends to coil the tube. When the pressure in the line increases, the tube straightens out slightly and the free end moves away from the center. This movement operates the linkage system and moves the pointer to a higher reading on the dial. When the pressure in the line decreases, the free end of the Bourdon tube moves toward the center and moves the pointer to a lower reading on the dial.

The gauge should be tested semiannually, or whenever there is any reason to doubt its accuracy. A gauge cock or shutoff valve will be installed between the pressure gauge and the pressure source so that the pressure gauge can be removed or repaired without relieving the pressure in the boiler.

A pressure gauge must be accurately calibrated and have a loop of tubing called a siphon (or pigtail) between the boiler and the gauge. (See Figure 11-2.) The exposed (uninsulated) coil is provided in the line leading to the gauge, and the steam condenses into water in this exposed coil. Thus there is always a condensate seal between the gauge and steam source. Sometimes the condensate seal is lost when repairs are made to the gauge or to the gauge line; in this event, the loop must always be filled with water before the gauge is subjected to steam line pressure.

# FUSIBLE PLUGS

Fusible plugs are used on some boilers for added protection against low water conditions. They are constructed of brass or bronze with a tapered hole drilled lengthwise through the plug. They have an even taper from end to end. This tapered hole is filled with a low-melting alloy consisting mostly of tin. There are two types of fusible plugs: (1) fire actuated, and (2) steam actuated.

## Fire-Actuated Type

The fire-actuated type is filled with an alloy of tin, copper, and lead, with a melting point of 445° to 450°F. It is screwed into the shell at the lowest permissible water level. One side of the plug is in contact with the fire or hot flue gases, and the other side is in contact with the water. As long as the plug is covered with water, the tin does not melt. If the water level drops below the plug, the tin melts and is blown out. The boiler then must be taken out of service to replace the plug.

## Steam-Actuated Type

The steam-actuated type of fusible plug is installed on the end of a pipe outside the drum. The other end of the pipe, which is open, is at the lowest permissible water level in the steam drum. A valve is usually installed between the drum and the plug. The metal in the plug melts at a temperature below that of the steam in the boiler. The pipe is small enough to prevent water from circulating in it. The water around the plug is much cooler than the water in the boiler as long as the end of the pipe is below the water level. However, if the water level drops below the open end of the pipe, the cool water runs out of the pipe and the steam then enters and heats the plug. The hot steam melts the tin and it is blown out by the steam, warning the operator. This type of plug can be replaced by closing the valve in the plug line. It is not necessary to take the boiler out of service to replace the plug.

Fusible plugs should be renewed once a year. *Do not* refill old casings with new tin alloy and use again. *Always use a new plug.*

## SAFETY RELIEF VALVE

The safety relief valve is the most important of boiler fittings. It is designed to open automatically to prevent the pressure in the boiler from increasing beyond the safe-operating limit. It is installed, in a vertical position, directly to the steam space of the boiler. (See Figure 11–9.) Each boiler has at least one safety relief valve; if the boiler has more than 500 square feet of heating surface, two or more valves are required. There are several different types of safety relief valves in use, but all are designed to open completely (pop) at a specified pressure, and to remain open until a specified pressure drop (blowdown) has occurred. Safety valves must close tightly without chattering, and must remain tightly closed after seating.

It is important to understand the difference between boiler safety relief

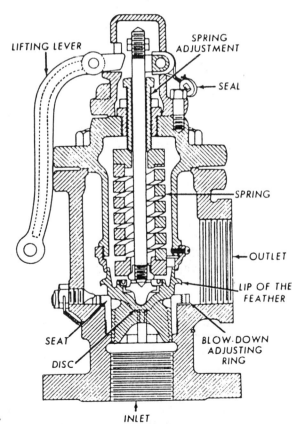

**FIGURE 11-9.**
Safety relief valve

valves and ordinary relief valves. The amount of pressure required to lift a relief valve increases as the valve lifts, because the resistance of the spring increases in proportion to the amount of compression. If a relief valve were installed on a steam drum, therefore, it would open slightly when the specified pressure was exceeded—a small amount of steam would be discharged—and then the valve would close again. Thus a relief valve on a steam drum would be constantly opening and closing, and this repeated action would pound the seat and disc and cause early failure of the valve. In order to overcome this difficulty, safety relief valves are designed to open completely at a specified pressure.

Several different types of safety relief valves are used on boilers, but they all lift on the same general principle. In each case the initial lift of the valve disc or feather is caused by static pressure of the steam acting upon the disc or feather. As soon as the valve begins to open, however, a projecting lip or ring of larger area is exposed for the steam pressure to act upon. The resulting increase in force overcomes the resistance of the spring, and the valve pops—that is, it opens quickly and completely. Because of the larger area now presented, the valve reseats at a pressure lower than that which caused it to lift originally.

Lifting levers are designed to lift the valve from its seat when boiler pressure is at least 75% of the pressure at which the valve is set to pop. Lifting levers are also used to check the action, and to blow away any dirt from the seat. When the lifting lever is used, the valve disc must be raised sufficiently to ensure that all foreign matter is blown from around the seat to prevent leakage after it is closed.

The various types of safety relief valves differ in the method of applying compression to the spring, the method of transmitting spring pressure to the feather or disc, the shape of the feather or disc, and the method of blowdown adjustment. Detailed information on the operation and maintenance of safety relief valves can be found in the instruction books which are furnished by the valve manufacturer.

## SURFACE BLOW VALVE

The surface blow valve, when installed, is located on the steam drum where it connects to the surface blow line which terminates at a scum pan inside the drum. This pan is located just below the normal water line. Its purpose is to remove scum from the surface of the water. As with most high-pressure valves of the globe type, the pressure is directed under the seat to facilitate opening.

## BOTTOM BLOW VALVES

Bottom blow valves, installed at the bottom of each water drum and header, are used to blow down the boiler, removing scale and other foreign matter that has settled in the lowest part of the water spaces. Boilers arc also blown down to control the concentration of solids dissolved in boiler water. Blowdown valves are located at the lowest point of the water sides.

## AIR COCK

The air cock is a valve located at the highest point of the steam space. Two important purposes of this valve are (1) to allow air to escape while raising steam on a cold boiler or when filling a boiler with water; and (2) to allow air to enter when draining the boiler. When lighting off a cold boiler, the air cock is one of the first valves to open. It is closed when sufficient steam has been generated to displace all the air in the steam space.

## ROOT VALVE

The root valve is an emergency shutoff valve located between the source of pressure and the supplied equipment. An example is the valve located between the main steam stop valve and the main steam line. This valve must be capable of withstanding the maximum operating pressure.

## REVIEW QUESTIONS

1. What type of control measures the steam pressure and controls operation of the boiler burner?
2. How is the pigtail between a control or gauge on a boiler installed?

3. What must be done before subjecting a control or gauge to steam pressure?

4. How many gauge glasses are required on a boiler?

5. What is the lowest water level permitted in horizontal fire-tube boilers?

6. What is the purpose of gauge cocks?

7. What device protects the boiler against low water by stopping the burners?

8. What should be done to electrode probe-type water level controls to prevent accidental shorting of the low water cutoff electrode?

9. Name the two types of fusible plugs used in boilers.

10. What device is used to prevent the boiler pressure from increasing beyond the safe-operating limit?

# CHAPTER 12

# HOT WATER CONTROLS

## Learning Objectives

The objectives of this chapter are to:

- Describe different types of controls used on hot water boiler systems
- Introduce the reader to the proper installation practices used when installing hot water boiler controls
- Show the operation of the different types of hot water controls used on hot water boilers
- Discuss different safety type controls used on hot water boilers.

Hot water boilers require a different type of control system than steam-type boilers. The controls used on these systems sense temperature rather than pressure. Hot water systems do not have the condensate return problems that are encountered with steam-heating systems. The hot water is circulated with a pump, sometimes called a circulator, which moves the water through the system of piping, coils, and valves. The boiler operator should be just as familiar with the controls used on these systems as he is with the ones used on steam applications.

## OPERATING TEMPERATURE CONTROL

This is the main temperature control used on hot water boilers. It is equipped with an immersion-sensing bulb that fits into a well inside the water pipe. (See Figure 12–1.) It is an adjustable control which is equipped with a set of electrical contacts that close on a rise in temperature. The contacts control the electrical circuit to the fuel valve. The sensing bulb is installed in a pipe at or near the water outlet of the boiler.

In operation, when the boiler water temperature drops to a predetermined setting the sensing bulb will cause the set of contacts to close. Electrical energy is then directed to the fuel valve and the boiler starts heating the water. When the water temperture has risen to the desired cutoff temperture, the sensing bulb will open the set of contacts, interrupting the electric current to the fuel valve. The boiler stops heating the water. The cycle is then repeated.

## HIGH-TEMPERATURE LIMIT CONTROL

The high-temperature limit control operation is much the same as the operating temperature control. It has an immersion-sensing bulb which opens a set of contacts on a rise in temperature. Some of these controls have two sets of contacts; one set that opens on a rise in temperature and another set that closes on a rise in temperture. Some are equipped with a manual lockout to prevent automatic operation during an overheated condition. (See Figure 12–2.) This is a safety control which operates only when an overheated condition occurs inside the boiler. The sensing bulb is installed in a pipe at or near the water outlet of the boiler.

In operation, when the water temperature inside the boiler has risen to

**FIGURE 12-1.**
Operating temperature control (Courtesy of Johnson Controls, Inc., Control Products Div.)

a dangerously high point and the operating control has not caused the fuel valve to close, the high-temperature limit control contacts will open the electrical circuit to the fuel valve, causing it to close. If a second set of contacts are included, an alarm will sound to warn the operator that a dangerous condition exists and needs attention. Operation of the boiler will not resume until the control is manually reset. The cause of the problem should be found and corrected or the problem will recur on the next heating cycle.

## INDOOR-OUTDOOR CONTROLS

Indoor-outdoor controls are temperature-sensing controls that have two sensing elements and two capillaries. One element senses the outdoor temperature and one senses the supply water temperature. (See Figure 12-3.)

A19

**FIGURE 12-2.**
High-temperature limit control wiring connections (Courtesy of Johnson Controls, Inc., Control Products Div.)

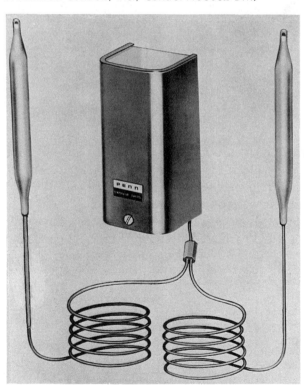

**FIGURE 12-3.**
Indoor-outdoor control (Courtesy of Johnson Controls, Inc., Control Products Div.)

The purpose of these controls is to either increase or decrease the boiler water temperature in response to the outdoor temperature. When the outdoor temperature drops, the control will raise the water temperature a certain number of degrees to compensate for the colder outdoor conditions. When the outdoor temperature rises, the control will reduce the supply water temperature and thereby reduce the operating costs.

## LOW WATER CUTOFF

The low water cutoff is a control which is mounted on the boiler at the lowest water level permissible to prevent damage to the boiler from overheating. These controls are float operated. They are equipped with either one or two sets of contacts, depending on the design of the equipment. One set will open when the water level reaches the predetermined level to close the fuel valve. The other set will close and sound an alarm to warn the operator that a problem exists. The float is installed at the lowest level permissible. (See Figure 12–4.) Should the water level drop to this point, onc

**FIGURE 12-4.**
Float-operated low water cutoff (Courtesy of McDonnell & Miller, ITT Fluid Handling Division)

set of the contacts will open the electrical circuit to the fuel valve. The other set will sound the alarm warning the operator of the problem. The boiler will need to be refilled with water before operation can be resumed. However, the reason for the low water must be found and corrected or the condition will recur.

## AIR THERMOSTAT

The air thermostat is a temperature-sensing device that places the boiler in operation in response to the outdoor air temperature. The sensing bulb is located outdoors. The control is set to start operation when the outdoor temperature drops to a predetermined point. (See Figure 12–5.)

**FIGURE 12-5.**
Air thermostat (Courtesy of Johnson Controls, Inc., Control Products Div.)

# REDUCING VALVE

The reducing valve is installed in the city make-up water supply to the boiler. It is set to a pressure that will force the water to the highest point in the system. (See Figure 12–6.) When determining what pressure to set the valve to, remember that 1 pound of pressure will cause water to rise approximately 3 1/3 feet. However, the water pressure should never be set greater than the recommended pressure for the boiler or the system.

# CHECK VALVE

The purpose of a check valve is to prevent boiler water from backing up into the city water main in case the boiler pressure should become greater than the city water main pressure. The check valve is installed in the city water line to the boiler. It should not be installed between the boiler and the relief valve if it is installed in this line. Check valves permit the water to flow into the boiler but prevent it from flowing from the boiler. (See Figure 12–7.)

# COMBINATION PRESSURE-TEMPERATURE GAUGE

The combination pressure-temperature gauge is mounted near the water outlet of the boiler. Its purpose is to sense the pressure and temperature inside the water system and to show these readings on a dial. This allows the operator to determine these conditions with only a glance.

# RELIEF VALVE

This is probably the most important water control on the boiler. Its purpose is to release the pressure inside the boiler to prevent an explosion or rupturing of the boiler when excessive temperatures exist in the system. (See Figure 12–8.)

A relief valve is a safety device that will open and relieve the pressure inside the system if all of the other controls should fail to operate as de-

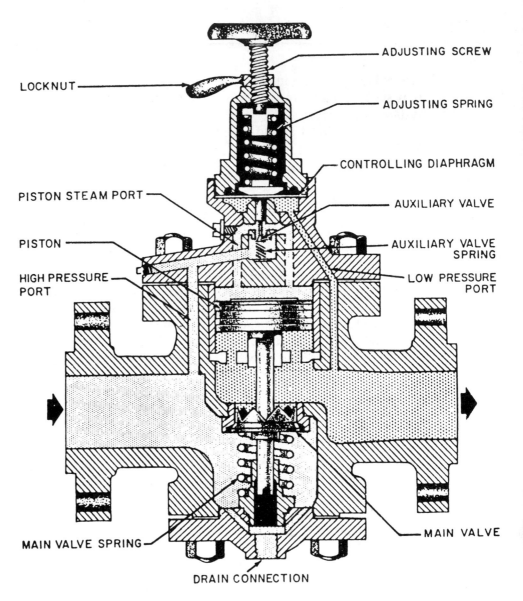

**FIGURE 12-6.**
Reducing valve

signed. When the relief pressure is reached, the valve will open and release the water from the system until the pressure has dropped to the point that the valve will reseat. The cause of this condition must be found and corrected before the boiler is placed back into operation.

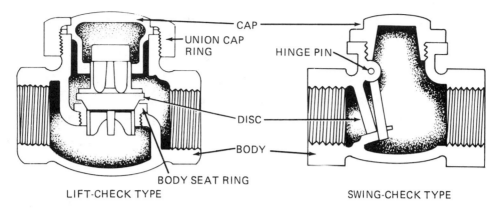

**FIGURE 12-7.**
Check valves

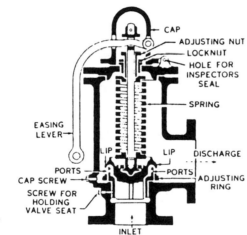

**FIGURE 12-8.**
Relief valve

# REVIEW QUESTIONS

1. What is the main temperature control used on hot water boilers?
2. Are the contacts in an operating temperature control normally open or closed?
3. What is the purpose of the high-temperature control on a boiler?
4. In the high-temperature limit control, what is the purpose of the second set of contacts?

5. What control is used to increase or decrease boiler water temperature in response to the outdoor temperature?

6. What control is mounted at the lowest permissible water level?

7. Where is the sensing bulb located on the air thermostat?

8. At what pressure is the reducing valve set?

9. What is the maximum pressure setting for a reducing valve?

10. What control is used to sense the pressure and temperature inside the water system?

# CHAPTER 13

# BOILER OPERATION

## Learning Objectives

The objectives of this chapter are to:

- Describe different procedures used in the day-to-day operation of a boiler
- Introduce the reader to the different uses of boiler operating logs
- Familiarize the reader with the preoperating checks made on boilers
- Show procedures used for lining up a system
- Introduce the reader to the operating procedures used in a boiler room.

It is a boiler operator's responsibility to operate and maintain boilers. In connection with the operation and maintenance of boilers, a boiler operator must work certain shifts, keep boiler operating logs, perform operator's maintenance on the boiler, perform preventive maintenance and minor repairs on boilers and associated equipment, complete chemical tests on boiler water and feed water, and carry out any other duties assigned by the boiler room manager.

# BOILER OPERATING LOGS

The main purpose of boiler operating logs is to provide a means of recording continuous data on boiler plant performance. Logs become a source of information that is needed for making an analysis of boiler operation for maintenance and repair purposes. The daily operating log sheets provide the basic information around which maintenance programming is developed. The log is arranged for a given period of time that is suitable for the particular installation. Log sheets may vary but you should have little difficulty in making log entries if you understand the types of information required. Some of the types of information that need to be entered in the log are listed below.

*Steam Pressure:*   The steam pressure entries are based on steam gauge readings and indicate the performance of the boiler.

*Steam Flow:*   Record the actual output of the plant in pounds per hour to obtain the steam flow rate. The data from these entries are used to determine the number of boilers to operate for the greatest efficiency.

*Feed Water Heater Pressure:*   The feed water heater pressure indicates if the proper operating temperature can be maintained in the heater.

*Feed Water Heater Temperature:*   The feed water heater temperature shows the effectiveness of the feed water heater. A drop in steam-supply pressure or insufficient venting may cause a low heater temperature.

*Feed Pump Pressure:*   This entry indicates the effectiveness of the boiler feed pumps. If the feed water supply fails, the pressure reading enables the operator to determine if the trouble is in the feed pumps. The pumps are defective if the feed pump pressure reading is below normal.

*Last-Pass Draft:*   This entry indicates the actual draft produced by the stack or the induced-draft fan. Ensure that you are familiar with the last-pass draft at various firing rates when the boiler is operating satisfactorily. A decrease in the last-pass draft with other conditions constant indicates leaking baffles. An increase shows that gas passages are be-coming clogged.

*Percent CO₂ Flue Gas:*   This value is a measure of the relative quantities of air supplied with the fuel. It is kept at a value which has been es-tablished as most satisfactory for the plant, the fuel, the firing rate, and other related factors. In plants not equipped with $CO_2$ recording meters, this value is determined with a hand gas analyzer. With expe-rience, the correct amount of air supplied to a boiler firebox can be determined from draft gauges and by observation. In all cases, find these values by using a hand gas analyzer

*Flue Gas Temperature:*   This is an indication of the quantity of heat leaving the boiler with the flue gases. This heat represents a direct energy loss in fuel. Abnormally high flue gas temperatures at given boiler firing rates are caused by dirty heating surfaces or leakage of baffles. If the heating surface has a coating of soot, heat which cannot escape is dis-charged to the stack. Leakage of heat through baffles allows gases to take a shorter path than intended, and reduces the contact of the gases with the entire heating surface. Excessive fouling of the firesides of boilers increases draft loss while leaking baffles decrease draft loss. Either condition raises the temperature of the flue gases above normal.

*Fuel:*   Always determine the quantity of fuel being used as this represents a major operating cost. Fuel oil quantities are determined by use of a measuring stick, in conjunction with tables supplied with a given tank. Some tanks are equipped with gauges that indicate the fuel volume. Gas quantities are determined by reading the gas meter.

*Outside Temperature:*   A heating plant load is greatly influenced by the out-side temperature. Record this temperature for comparison with the amount of steam generated and the amount of fuel used. These com-parative values are useful in determining abnormal fuel consumption and in estimating future requirements.

*Make-Up Water:*   Record the quantity of make-up water used, to enable the operator to note an abnormal increase before a dangerous condi-tion develops. Return all possible condensate to the boiler plant in or-der to save water and the chemicals used to treat the water.

*Water Pressure:*   The feed water pressure is most important to safe boiler-plant operation. A record of water pressure indicates whether or not the water is sufficient.

*Hot Water Supply Temperature:*   Record the hot water supply temperature. Insufficiently heated water can cause scaling or deposits in a boiler.

*Water Softeners:*   Where softeners are used, keep a meter record to inform the operator when units must be regenerated. A decrease in the quantity of time used for runs between regeneration indicates either an increase in hardness of the incoming water or deterioration of the softening material.

*Total and Averages:*   Space is provided on some charts for recording the total and average quantities per shift.

*Steam Flow:*   The steam-flow meter integrator reading at the start of a shift, subtracted from the reading at the end of a shift, and multiplied by the meter constant gives the quantity of steam generated. Dividing the steam generated by the fuel burned yields a quantity which indicates the economy obtained. If a plant does not have a steam-flow meter, pumps may be calibrated for flow and a record kept of their operating time, or condensate and make-up water may be metered.

*Boiler Feed Pumps in Service:*   A record of the boiler feed pumps in service makes it possible to determine the operating hours and to ensure that various pumps are used for equal lengths of service.

*Phosphate, Caustic Soda, and Tannin Added:*   A record of the phosphate, caustic soda, and tannin used is valuable in keeping the correct boiler water analysis and in determining the total chemicals used.

*Remarks:*   The remarks column is used to record various types of information for which space is not provided elsewhere on the log sheet. List the equipment to be checked daily. Note all irregularities which are found in connection with all inspections. List the date when boilers are drained and washed out thoroughly at intervals determined by local water conditions. Indicate the condition of internal cleaniness.

*Other Personnel:*   The names of the personnel responsible for specific tasks and data must be entered on the log sheet if required.

## BOILER OPERATION

The operation of a boiler consists of six major phases: (1) preshift assumption checks, (2) preoperating checks, (3) lining-up systems, (4) operating procedures, (5) operating checks, and (6) securing procedures.

### Preshift Assumption Checks

The preshift assumption checks are often neglected by boiler shift-standers. Before you assume the responsibility of boiler shiftstander, you should complete specified checking procedures to ensure that the equipment is in sound operating condition and is functioning satisfactorily. Normally, when the shift is relieved, the shiftstander coming on duty inspects the instrument readings and charts, visually inspects all equipment, and exchanges information with the off-going shiftstanders. On-coming shiftstanders should complete the following inspections and tests before assuming duty:

1. Visually inspect the setting and the casing.
2. Observe the furnace and firing conditions.
3. Inspect all charts to determine apparent performance of the equipment, controls, etc.
4. Inspect all fans, dampers, damper drives, and other driven auxiliaries.
5. Test the water columns and gauge glasses.
6. Test the alarm units.
7. Obtain information regarding the boiler's operating condition and ask the shiftstanders on duty if any unusual events or troubles occurred during their shift period.

Immediately after assuming operational responsibility, a complete inspection of all auxiliary equipment should be made, as follows:

1. Inspect all electric motor drives for evidence of abnormal temperature, condition of bearings, etc.
2. Inspect the fan and pump bearings for evidence of overheating and adequacy of lubrication.
3. Visually inspect the boiler and all associated equipment, being careful to listen for unusual sounds, friction, vibration, and other abnormal conditions.
4. Inspect the burners, fuel supply, pilot systems, and other fuel supply components.
5. Review the log sheets to obtain information on past operating conditions and unusual events that occurred.

## Preoperating Checks

The preoperating checks should be completed before lining-up and lighting-off a boiler. These checks are performed to ensure that the plant and associated equipment are in safe and efficient operating condition. The major preoperating procedures applicable to boilers in general are described below.

1. Inspect the boiler room or area for safety and fire hazards.
    a. Remove rags, paint cans, oil spots on the floor, and so on.
    b. Remove tools and equipment that may cause injury from the area.
2. Furnace and gas passages
    a. Must be clean and clear
    b. Must have doors that fit tight
    c. Must be in good repair
    d. Should have no oil or tools in the combustion chamber
    e. Must be purged
3. All valves must be inspected for
    a. Good operating condition
    b. Bent stems
    c. Missing or broken handwheels
4. Piping
    a. Must be inspected for leaks
    b. Must be checked for proper supports
5. Electrical systems should be checked for
    a. Oil-soaked or frayed wire insulation
    b. Damaged or loose conduit
    c. Improperly secured control boxes
6. Guards for moving parts
    a. Should be tight and in proper position
7. Water gauge glass
    a. Must be well lighted
    b. Must not be stained
    c. Must not leak at connection
8. Dampers
    a. Must not stick
    b. Must be in good working order
    c. Must be open before lighting off
9. Air cock
    a. Must be opened and water level on cold boiler must be adjusted
    b. Must be opened to vent boiler of air, if boiler is cold
10. Pressure gauge
    a. Must be correct and clean

b.   Must be well lighted

c.   Cock in line must be open

11. Steam drain lines

a.   Must be opened if line is cold

12. Protective controls

a.   Must be checked for low water cut-off

b.   Must be tested frequently for flame failure with burner operating; frequency depends upon equipment requirements

13. Ensure that the following auxiliary equipment is in safe and operating condition and is lubricated:

a.   Boiler feed pumps

b.   Oil pumps

c.   Draft fans

d.   Feed tank

e.   Feed water heater

**Gas-Fired Boilers.**   Before operating gas-fired boilers the following items must be checked, in addition to those listed above:

1. The pilot and main gas cock should be operating smoothly.

2. All copper tubing must be free from restriction due to kinks and flat spots.

3. Air shutters must operate freely; linkage must not have an excessive amount of lost motion.

4. Burner and main gas valve must be firmly supported.

5. Be sure that the boiler room is free of gas. If gas is present, ventilate and test piping with soap solution.

**Oil-Fired Boilers.**   Before operating oil-fired boilers the following items should be checked:

1. Strainers

a.   Inspect and clean

b.   Renew if wire mesh is defective

2. Burners

a.   Inspect and clean if necessary

b.   Clean nozzle if necessary

c.   Inspect and set electrodes

   d.  Check that fittings do not leak
   e.  Check operation of burner safety switch
3. Oil system
   a.  Inspect for leaks; make necessary repairs if leaks are present

## Lining-up Systems

After the preoperating checks have been completed, the next thing to do is line up the boiler systems—fuel, water, steam, and electrical equipment. These procedures will vary with the different types and kinds of boilers. The manufacturer's recommendations should always be followed.

Before lining up a boiler, complete the following basic tasks:

1. Fuel oil
   a.  Measure with a stick or gauge
   b.  See that the proper valves are open
   c.  Remove any excess accumulation of water in the fuel tank
2. Gas
   a.  Check the gas pressure
   b.  Check for gas leaks
3. Gas-fired unit
   a.  Check and regulate the water level; line up the feed system
   b.  Examine the burner, control valves, and safety cutouts for proper working condition before lighting off the boiler
   c.  Purge all air out of the gas lines by external vents before lighting off the boiler
   d.  Check the draft devices and purge the combustion chamber
   e.  Light the pilot and set the flame
   f.  Open the main gas cock
   g.  Close the burner control switches which will ignite the burner
   h.  Maintain the fuel-air ratio to maintain complete combustion
4. Oil-fired unit
   a.  Check and regulate the water level; line up the feed water system; check the feed pump operation
   b.  Line up the fuel oil system
   c.  Purge the combustion chamber
   d.  Close the burner control switch; if automatic, the burner should light off

e.  Should the ignition attempt fail, purge the furnace before the second attempt

f.  Prevent the fuel oil from impinging on the brickwork or part of the boiler

g.  Maintain the proper air-fuel ratio.

The lighting-off procedures listed below are applicable, in general, to most boilers.

1.  Close the following valves:
    a.  All blowdown valves
    b.  Boiler drains
    c.  Chemical feed valves
    d.  Boiler nonreturn valves
    e.  Main-steam-stop valve
    f.  Soot blower header (steam system)
    g.  All burner fuel valves
    h.  All soot blowers
    i.  Water column drain valves
    j.  Feed water regulator drains
    k.  Auxiliary valves, as necessary

2.  Open the following valves:
    a.  Vent valves on boiler drums and superheaters
    b.  Superheater drain valves
    c.  Recirculating line valves in economizer, if so fitted
    d.  Feed water stop and check
    e.  Drum steam gauge connection
    f.  Water column gauge connections
    g.  Water column gauge glass valves
    h.  Auxiliary valves, as necessary

3.  Start filling the boiler with properly treated water at a temperature relatively close to the temperature of the pressure parts. The temperature difference should not be greater than 50°F to avoid severe temperature stresses. Fill the boiler to a level just below the middle of the glass on the water column.

4.  Close the induced-draft fan dampers (or other flue gas control dampers).

5.  Start the induced-draft fan.

6.  Close the forced-draft fan dampers (or other air control dampers).

7. Start the forced-draft fan.

8. Start the air heater rotor, if a regenerative-type air heater is installed.

9. Light-off the boiler in accordance with the manufacturer's instructions and maintain a firing rate so that the water temperature in the boiler is raised at the rate of 100°F per hour until the operating pressure is reached. On new boilers, expansion movement should be checked to see that no binding or interference occurs.

10. When burning oil, prevent incomplete combustion in the boiler furnace; any unburned oil will be deposited on the cooler surfaces in the back of the unit, such as the economizer and air heater. This creates a potentially dangerous situation.

11. When the steam drum reaches approximately 25 psig, close the vent valves on the boiler drum. Check the steam pressure gauge at this time to be sure that it is registering.

12. Ease up on the stem of the main-steam-stop valve to prevent any serious expansion stresses. If there is no steam on either side of the main-steam-stop valve, gently lift and reseat it to make sure that it is not stuck. Open the drain valve on the boiler side of the main-steam-stop valve.

13. Observe the water level carefully, to ensure that no water is carried over into the superheater. Maintain a normal water level in the drum by blowing down or feeding water as may be required.

14. Operate the vent and drain valves in the superheater headers and the economizer by following the manufacturer's instruction. In general, drain valves in the superheater inlet header are closed first, followed by the drains in the superheater outlet header. In any case, the superheater outlet header drain and vent valves must not be completely closed until an adequate steam flow through the boiler outlet valve is assured.

15. Check for leaking gasket joints. If a leaking gasket is discovered, shut down the boiler and tighten the joints.

16. If the gasket still leaks, drop the pressure again, replace the gasket and repeat the lighting-off sequence.

Before cutting-in the boiler, proceed as follows:

1. Open all drain valves between the boiler and the header, especially the drains between the two stop valves.

2. Warm up the steam line between the boiler and the header by backfeed through the drip line or by means of the bypass valve.

3. When the steam line is thoroughly heated and at header pressure, open the header valve.

4. When the boiler pressure almost reaches line pressure, open the bypass line around the main steam stop valve to equalize the pressures and temperatures in the piping; then slowly open the main-steam-stop valve. As the boiler reaches line pressure and is actually steaming, the nonreturn valve stem is slowly raised to the full open position.

5. After the boiler is on the line, close all superheater drains.

6. Inspect the entire boiler, and close any drain valves that are not discharging condensate.

7. Close the economizer recirculating valve, when an adequate continuous feed water flow is established.

8. Close the drain valve at the nonreturn valve.

9. Close the bypass valve around the nonreturn valve.

10. When operating a boiler having a pendant (nondrainable) superheater, the operation will be slightly different. Superheaters of this type will trap condensate in the loops which must be boiled off before the firing rate can be increased and the steam flow is started.

11. It is very important to maintain a constant firing rate. The strength of thick drums may be impaired by excessive temperature differentials between the top and the bottom of the drum, if the proper firing rate is not maintained. Tubes may start leaking at the rolled seats and the superheater tubes may overheat.

12. On boilers generating saturated steam, it is necessary to follow the manufacturer's specified firing rate and follow the above instructions for removing air and condensate.

## Operating Procedures

The successful operation of boilers depends largely upon the operator being well informed and constantly vigilant. No fixed set of rules can be established that will be adequate for all conditions. Consequently, the operator must see and interpret all prevailing operating conditions and, if necessary, take action to control, modify, or correct them. To be able to do this, the operator must be thoroughly familiar with the characteristics and standard operating procedures for the boiler for which he is responsible. This section will acquaint you with the basic operating procedures which will apply to most, if not all, boilers that you will operate. For specific operating instructions, consult the manufacturer's manual for the boiler concerned.

**Normal Operation.**  During normal boiler operation, the operators have two major responsibilities. The first is to see that the proper water level is carried at all times. If the water level is too low, tubes may overheat, blister and rupture. If the water level is too high, carry-over of water to the superheater tubes may occur, causing damage to the superheater elements and to the turbine. Check the water gauge frequently to be certain that it is reading correctly and that the proper water level is being maintained.

The second major responsibility is to prevent loss of ignition when burning fuel is in suspension, as in the case of gas or oil. Maintain safe and efficient combustion conditions in the boiler furnace and the correct fuel-air flow ratios.

**Blowdown.**  Establish definite intervals for blowing down the boiler, depending on the type of operation and the chemical analysis of the boiler water. During regular operation, never blowdown economizers or water-cooled furnace walls.

Blowdown valves on this type of equipment are provided to serve only as drain valves. When using low-point drains or blowdown valves, boilers should be blown down at reduced or moderate rates of steaming. Where the water glass is not in full view of the operator blowing down a boiler, another operator should be temporarily assigned to observe the water glass and signal the operator manipulating the valves. For control of water conditions when working, it is wise to use a continuous blowdown to maintain the proper concentration at all times, and to prevent blowing down large quantities of water while the boiler is operating at a high capacity.

**Boiler Makeup.**  Use only properly treated water for make-up purposes, and maintain the boiler water conditions as specified in the water treatment instructions. Make an accurate water analysis at specified intervals. Carefully control the blowdown and the addition of treatment chemicals to meet the manufacturer's specifications.

**Soot Removal.**  Remove soot from the hoppers and pits at definitely established intervals, as necessary.

**Instrument Readings.** Establish definite intervals for observing and recording the readings on all important instruments and controls. Be sure you obtain accurate readings and see that the readings are recorded properly on the log sheet or other required record.

## Operating Checks

To help ensure efficient operation of the boiler, see that the following operating checks are carefully made:

1. Water level
   a. Check frequently as water expands during a heating up period
2. Main-steam-stop bypass (if installed)
   a. Open if boiler is to be cut in on a cold line
   b. Main stop can be opened if no other boiler is on the same steam line
3. Air cock
   a. Close after the steam has formed and has blown all air from the boiler
4. Steam pressure
   a. Raise slowly—usually 1/2 to 2 1/2 hours, depending on type, size, and condition of boiler
   b. Temperature of water should be raised at a rate of 100°F per hour
5. Safety valves
   a. Manually lift when the pressure is at least 75% of the valve setting
   b. Make sure that the valves seat properly; if the valves fail to reseat properly, lift a second time
6. Cutting boiler in
   a. If closed, open the main-steam-stop valve slowly to avoid thermal shock; avoid water hammer
7. Boiler feed water
   a. Commence feeding boiler as needed
   b. Probably will be automatically controlled
8. Gas firing
   a. Maintain ignition

    b.   There should be no soot formation

    c.   Maintain proper air-fuel ratio

9. Oil firing
    a.  Maintain ignition
    b.  Observe flame and adjust dampers; check accuracy by flue-gas analysis

10. Water level
    a.  The first duty when taking over the watch is to blowdown the gauge glass and water column. (NOTE: Observe the promptness of the return of water into the glass.)
    b.  Keep at proper level
    c.  Do not depend upon automatic regulators
    d.  At frequent intervals compare readings of different methods to determine true water level; use try cocks

11. Boiler blowdown
    a.  Rids boilers of mud, scale, rust, and other suspended impurities in the water
    b.  Open quick-covering valve first; slow opening second. Close in reverse order.
    c.  Frequency and amount of blowdown depends on water tests
    d.  Keep eye on gauge glass

12. Efficient operation
    a.  Keep flue gas temperature low; it should be about 150°F higher than the temperature of the steam being generated
    b.  Take flue gas analysis periodically; maintain the proper $CO_2$ level for the fuel used
    c.  Flame should be long and lazy; must not enter tubes; must not be dark and smoky; should have light brown haze from stack, except gas; should have a yellow flame with dark or almost smoky tips when fuel is oil
    d.  Draft should usually be 0.03 to 0.06 inch of water column; extra is needed when the stack and refractory are cold. (Check the manufacturer's recommendations.)
    e.  Maintain a low make-up feed rate; repair leaky steam and return lines; avoid excessive boiler blowdown
    f.  Keep the boiler and lines well insulated
    g.  Maintain proper feed water temperature
    h.  Carry out prescribed treatment of boiler water

## Securing the Boiler

The recommended procedures for securing a boiler are as follows:

1. Reduce the load on the boiler slowly, cutting out the fuel supply by proper operation of the fuel burning equipment.
2. Maintain the normal water level.
3. When the boiler load has been reduced to 20% of the normal firing rate, change the combustion control and the feed water control to manual operation.
4. Before securing the final fuel burner, open the drain valves at the steam and nonreturn valve, and the drain valve on the superheater outlet header. Be sure that the bypass valve around the nonreturn valve is closed.
5. Secure the final fuel burner when the load has been reduced sufficiently.
6. Continue operation of the draft fans until the boiler and the furnace have been completely purged.
7. Shut down the draft fans.
8. Close the dampers, including the air heater and superheater bypass dampers, when provided.
9. Follow the manufacturer's instructions for the rate of cooling the boiler. A thermal strain may occur if the change is too fast.
10. When the boiler pressure has started to drop, close the steam stop valve and nonreturn valve.
11. When the boiler no longer requires any feed and nonreturn valve is closed, open the valve in the recirculating connection of the economizer, if provided.
12. Let the boiler pressure drop by relieving the steam through the superheater drain valve and through the drain valve at the nonreturn valve. If the boiler is losing pressure at a rate faster than specified by the manufacturer, throttle the drain valves as necessary to obtain the proper rate. Do not close the valves completely.
13. When the drum pressure drops to 25 psig, open the drum vent valves.
14. If a regenerative-type air heater is used, the rotor may be stopped when the boiler exit gas temperature is reduced to 200°F.

15. The boiler can be emptied when the temperature of the water in the boiler is below 200°F. Before sending personnel into any part of the boiler, close and properly tag all controls, valves, and drains or blowdown valves connected with similar parts, or other units under pressure at the time. This will prevent any steam or hot water from entering the unit. The tags are to be removed only by an authorized person and must remain in place until all personnel have completed their work. Ventilate the boiler thoroughly and station a man outside. Inside, use only low-voltage portable lamps with suitable insulation and guards. Even 110 volts can be lethal under the conduction conditions found inside a boiler. All portable electrical equipment should be grounded, and electrical extensions should be well-insulated, designed to withstand rough usage, and maintained in good condition.

## Boiler Emergencies

Typical emergency situations encountered in connection with the operation of boilers are

1. Low water
2. High water
3. Serious tube failure, making it impossible to maintain water level
4. Flareback caused by an explosion within the combustion chamber
5. Minor tube failure indicated by trouble maintaining water level under normal steam demand
6. Broken gauge glass on a water column

Table 13–1 gives the safety procedures to follow when these boiler emergencies occur.

# SAFETY

When it comes to the duties involving the operation or servicing of boilers, the need for safety cannot be overemphasized. Much progress has been made over the years in the development of safety devices for boilers. There are

**TABLE 13-1**

Boiler Emergencies

| Tasks | Key Points |
|---|---|
| EMERGENCY ONE: LOW WATER CONDITION INDICATED BY NO WATER LEVEL IN THE GAUGE GLASS.<br>Secure the boiler, secure electrical switches, steam stop and feed stop. Prove water level by opening try cocks. Cool the boiler slowly until the water temperature is 200° F. Secure all sources of draft. Check controls. Find out the cause for low water level. Correct the trouble. After correction has been made, add water to obtain the correct water level. | DON'T ADD WATER TO THE BOILER to raise water level in the gauge glass column. STAY AWAY from discharge. DON'T FORCE COOL. |
| EMERGENCY TWO: HIGH WATER CONDITION INDICATED BY GAUGE GLASS FULL OF WATER.<br>Prove water level by opening the try cocks. Blowdown boiler by opening blowdown valves. Find out the cause of high water condition. Check feed pump controls. Correct the trouble. Secure the boiler if pump controls operate improperly. | STAY AWAY from discharge. Check blowdown pit. Watch the gauge glass until normal level is reached. If control operates properly, continue to operate the boiler. |
| EMERGENCY THREE: SERIOUS TUBE FAILURE MAKING IT IMPOSSIBLE TO MAINTAIN WATER LEVEL.<br>Secure the boiler by securing the electrical, steam, and fired systems. Add water to the boiler until the ruptured tube level is reached and the boiler is cooled to a temperature of 200° F. Open the boiler to replace the tube. | For large boilers: Water should not be fed to boiler until properly cooled. Mark the gauge glass if within its range. Observe level by whatever means available. |
| EMERGENCY FOUR: FLAREBACK CAUSED BY AN EXPLOSION WITHIN THE COMBUSTION CHAMBER.<br>Secure the boiler. Find out the cause of flare-back and correct the trouble. Check for sufficient fuel and any type of fuel contamination. Check the burner. | Ensure that a slug of water did not interrupt flame with a refire before prepurge. |
| EMERGENCY FIVE: MINOR TUBE FAILURE INDICATED BY TROUBLE MAINTAINING WATER LEVEL UNDER NORMAL STEAM DEMAND.<br>Secure the boiler if it is possible to remove it from the line to allow sufficient time to make necessary repairs. Secure electric switches. Open steam stop and feed stop if additional water is not needed to protect remaining tubes. | If unable to secure due to steaming requirements, and you can maintain water level, continue to operate. If unable to maintain water level and/or supply, secure the boiler. |
| EMERGENCY SIX: BROKEN GAUGE GLASS ON WATER COLUMN.<br>Secure top and bottom valves immediately. Use chains or whatever method available to prevent injury to personnel. Replace the gauge glass. | Boiler may be kept on the line, if necessary. Check the boiler water level by using try cocks. |

still many ways, however, in which serious accidents on happen around boilers. A boiler operator who is careless in the performance of his job poses a threat not only to his own safety, but to the safety of others. It seems that accidents somehow have a way of happening at the least expected moment. This is why constant alertness and close attention to detail are imperative. Don't take chances! Be *safety conscious!*

Some of the major safety precautions to be observed by those engaged in boiler opearation and servicing are presented below.

As protection against toxic or explosive gases, boiler settings must be ventilated completely and tested for the presence of toxic or explosive gases before personnel are permitted to enter the boiler.

The covers of manholes must be removed for ventilation purposes before personnel enter the drum.

Before anyone enters the steam drums, mud drums, or other water-side enclosures, steam and feed lines connected to the headers under pressure should be isolated by a stop valve and a blank, with an open tell-tale valve in between.

A ventilating fan should be operated in the drum when a man is working in the boiler.

Workers should not be inside the water side of the boiler when pressure is being applied to a test valve which has not been under pressure.

Workers should wear protective clothing when making boiler water tests.

Boiler settings must be examined daily for external air leaks, cracks, blisters, or other dangerous conditions in joints, tubes, seams, or blowoff connections. Any problems are to be reported to the person in charge immediately.

Boilers should also be examined regularly for deposits on their heating surfaces and for grease or other foreign matter in the water. Boilers showing any such irregularities should be cleared at first opportunity and should not be used until cleared.

Performing certain adjustments and repairs while pressure is up is prohibited. A complete absence of pressure is to be ensured by opening the air cock or test and water gauge cocks that connect with the steam space. This is to be done before fittings or parts subject to pressure are removed or tightened, and before manhole or handhole plate-fittings are loosened on a boiler which has been under pressure.

Combustion control, feed control, and burner, stoker, or similar adjustments are permitted with the boiler steaming, recognizing that many adjustments can only be made when the steam pressure is up.

When performing cleaning operations, workers should wear the proper personal protective equipment. The following requirements apply:

1. Hard hats and goggles must be worn.
2. When a worker is chopping slag inside a furnace, a respirator must be worn.
3. Safety-toe shoes or toe guards must be worn to prevent injuries from falling slag.

When a man is working inside the furnace, a large warning sign, such as *Caution—Man Working Inside,* should be placed near the furnace entrance.

The use of open flame lights is prohibited in boilers. When cleaning where flammable vapors and gases may be present, workers are to use only explosion-proof portable lamps equipped with heavily insulated 3-wire conductors, with one conductor connecting the guard to the ground.

Any oil which has accumulated on furnace bottoms should be cleaned at least every 8 hours, and more frequently if necessary.

Condensate pits in boiler rooms should be provided with metal covers. Where it is necessary that such pits be open for maintenance, adequate guards should be placed around them and warning signs posted.

Wear goggles with dark lenses, number 1.5 to 3 shade, and suitable fireproof face shields when working near or looking through the furnace doors of boilers in operation.

When firing a cold boiler, be sure that the air vents are open on the boiler proper and that the drains are open on the superheater; these should be kept open until steam is liberated from the openings. Superheater vents must remain open until the boiler is on the line.

Be sure that gas-fired and oil-fired boilers, whether manual or automatic, are cleared of combustible gases after each false start.

All semiautomatic (multiburner) boilers and fully automatic boilers should be equipped with a manually activated switch for pilot ignition and a control device to prove the pilot flame before the main fuel valve is opened. *Do not use a hand held torch to light off a boiler.* If a hand held torch is applied to a firebox filled with vaporized oil, a severe boiler explosion is likely to result.

Prevent overheating of boilers equipped with superheaters by firing at a slow rate during the warm-up period and allowing a small amount of steam to flow through the superheater.

When taking over a shift, blow the water gauges and note the return of the water in the glass. Be certain of the water level at all times. *Do not* be misled by a dirt marking on the gauge which may look like the surface of the water. *Do not* depend entirely upon automatic alarm devices and automatic feed water regulators.

If the water goes out of sight in the bottom of the gauge glass, kill the fire, utilizing the quickest means available, immediately close the steam stop valve, and allow the boiler to cool slowly. Then drain the boiler completely and open for inspection. *Do not feed cold water into a boiler that has had low water until the boiler has cooled.*

Check the water on steaming boilers by try cocks at least once each shift and before connecting the boiler to the line. Remember that a fall in steam pressure may indicate low water.

Check safety valves at frequent intervals to be sure that they will pop at the correct pressure, as marked on the nameplate. Do not break the seal of a safety valve or change its adjustment unless such action has been authorized. *Never* weight pop valves, relief valves, or similar controls to increase the recommended steam pressure for which the boiler is approved.

Do not use oil from a tank in which a considerable amount of water is mixed with the oil unless a high suction connection is provided. When an atomizer sputters, shift the suction to the standby tank or another storage tank. A sputtering atomizer indicates water in the oil.

Minimize the fouling of oil heaters by using as few heaters as possible. Recirculate the oil through the used heaters for a short time after securing the burners. Maintain the prescribed fuel-oil temperature; *do not* exceed it.

If a large steam leak occurs in a boiler, shut off the burners, continue to feed water until the fire is out, close the steam stop valve, ease the safety valves, clear the furnace of gases, close the registers, and cool the boiler slowly.

Do not tighten a nut, bolt, or pipe thread, nor strike any part, nor attempt to make any other adjustments to parts while the boiler is under steam or air pressure.

Exercise precautions to prevent lubricating oil, soap, or other foreign substances from getting into the boiler. Condensate from cleaning vats should be drained to waste and not returned to the boiler.

Close the furnace openings as soon as all fires have been extinguished and the furnace has been cleared of all gases.

Do not use water to extinguish an oil fire in the furnace.

When the fires are banked, make certain that the draft is sufficient to carry off all flammable gas accumulations.

In case of an oil fire in the boiler room, close the master fuel-oil valve and stop the oil pump.

In addition to the above precautions, the following list contains a number of safe practices which you should endeavor to follow earnestly in your work. It also contains a number of unsafe practices which you should make a vigorous effort to avoid.

## Never

Never fail to anticipate emergencies. Don't wait until something happens before you start thinking.

Never start work on a new job without tracing every pipeline in the plant and learning the location and purpose of each and every valve regardless of size. Know your job!

Never leave an open blowdown valve unattended when the boiler is under pressure or has a fire in it. Play safe, memory can fail.

Never give verbal orders for important operations or report such operations verbally with no record. Have something to back you up when needed.

Never light a fire under a boiler without a double check on the water level. Many boilers have been ruined this way.

Never light a fire under a boiler without checking all valves. Why take a chance?

Never open a valve under pressure quickly. The sudden change in pressure, or resulting water hammer, may cause piping failure.

Never cut a boiler in on the line unless its pressure is within a few pounds of header pressure. Sudden stressing of a boiler under pressure is dangerous.

Never bring a boiler up to pressure without trying the safety valve. A boiler with its safety valve stuck is nearly as safe as playing with dynamite.

Never increase the setting of a safety valve without authority. Serious accidents have occurred from failure to observe this rule.

Never change adjustment of a safety valve more than 10%. Proper operation depends on the proper setting.

Never allow unauthorized persons to tamper with any steam-plant equipment. If they don't injure themselves, they may cause injury to you.

Never allow major repairs to a boiler without authorization.

Never attempt to light a burner without venting the furnace until clear. Burns are painful.

Never fail to report unusual behavior of a boiler or other equipment. It may be a warning of danger.

## Always

Always study every conceivable emergency and know exactly what moves to make.

Always proceed to the proper valves or switches rapidly but without confusion in time of emergency. You can think better walking than running.

Always check the water level in the gauge glass with the try cocks at least daily and at any other time if you doubt the accuracy of the glass indication.

Always accompany orders for important operations with a written memorandum. Use a log book to record every important fact or unusual occurrence.

Always have at least one gauge of water before lighting off. The level should be checked by the gauge cocks.

Always be sure blowdown valves are closed, and proper vents, water column valves, and pressure gauge cocks are open.

Always use the bypass if one is provided. Crack the valve from its seat slightly and await pressure equalization. Then open it slowly.

Always watch the steam gauge closely and be prepared to cut the boiler in, opening the stop valve only when the pressures are nearly equal.

Always lift the valve from its seat by the hand lever when the pressure reaches about three-quarters of popping pressure.

Always consult the person in charge of the boiler plant and accept his recommendations before increasing the safety-valve setting.

Always have the valve fitted with a new spring and restamped by the manufacturer for changes over 10%.

Always keep out loiterers, and place plant operations in the hands of the proper persons only. A boiler room is not a safe place for a club meeting.

Always consult the person in charge of the boiler plant before making any major repair to a boiler.

Always allow draft to clear the furnace of gas and dust for several minutes. Change the draft control slowly.

Always consult someone in authority. Two heads are better than one.

# REVIEW QUESTIONS

1. What is the purpose of boiler operating logs?
2. What does the temperature of the boiler flue gas represent?
3. Name the six major phases of boiler operation.
4. What is the purpose of the preshift assumption check?
5. What check is performed to ensure that the plant and associated equipment are in safe and efficient operating condition?
6. What temperature should the water be when filling a boiler?
7. What should the water level be in a steam boiler?
8. What is very important when lighting off a cold boiler?
9. Name the two major responsibilities of a boiler operator during normal operation.
10. When should economizers or water-cooled furnace walls never be blown down?
11. When is it best to blowdown a boiler with low-point drains or blowdown valves?
12. When should a boiler water analysis be made?
13. How often should the water level of an operating boiler be checked?
14. What is the purpose of blowing down a boiler?
15. When securing a boiler, at what point should it be changed to manual operation?
16. At what water temperature can a boiler be emptied?
17. Why should the covers of manholes be removed before entering the boiler for inspection or repairs?
18. What should be done before tightening or removing the pressure parts of a boiler?
19. What type of protection should be taken when looking through the furnace door of an operating boiler?
20. Should cold water be fed into a hot boiler?

# CHAPTER 14

# MANAGING BOILER OPERATION

## Learning Objectives

The objectives of this chapter are to:

- Introduce the reader to the proper procedures used in the management of boiler operation
- Show the need for safety training and practice
- Familiarize the reader with the procedures used in idle boiler care and lay-up
- Introduce the reader to the procedures used when taking a boiler off the line.

As you gain experience in the operation and maintenance of a boiler room, you will probably be called upon to serve as a crew leader of various teams which can consist of any number of workers. The teams will perform duties which are related to the maintenance, repair, and operation of the boiler room and its related equipment.

After this, you may be placed in complete charge of the boiler room and its operations, depending upon the size of the operation. The duties of a boiler room manager may vary from one installation to another. At most installations, however, a boiler room manager's duties involve planning work assignments, supervising work teams, initiating requisitions, keeping time cards, and reviewing the daily logs.

# PLANNING WORK ASSIGNMENTS

For purposes of this discussion, planning is the process of determining the requirements and devising and developing methods for completing a project. Proper planning saves time and money, and makes the job easier for everyone. Some pointers that will help you plan day-to-day work assignments are given below.

When you determine that a specific job is to be done, one of the first things that you should do is to understand clearly just what is to be done. Study the plans and specifications where applicable. If you have any questions, ask those in a position to know the answers to supply the information that you need. Among other things, make sure that the priority of the job is understood; find out the time frame for completion of the project; and ask if there are any specific instructions that are to be followed.

In planning for the project, you must consider the capabilities of the workers available for the assignment. Determine who is to do what and how long it should take to complete the job. Plan work assignments far enough in advance that you can have another job ready for each worker as soon as the project is completed.

Establish goals for each work day and encourage your people to work together as a team in accomplishing these goals. You want your goals to be such that your workers will be busy, but make sure your goals are realistic. During an emergency, most workers will make a tremendous effort to meet the deadline. But men are not machines, and when there is no emergency, they cannot be expected to continuously achieve a high rate of production. In your planning, you must allow for things which are not considered direct labor, such as safety training, disaster control training, and vacations.

To help ensure that the job is done properly and on time, you will want to consider which method to use in accomplishing the job. If there is more than one way of doing a particular job, make sure the method you select is the best way. After selecting a method, analyze it to see if it can be simplified with a resultant savings in time and effort.

Plan material requirements so that you will not have a lot of materials left over. But don't estimate your material so low that you might run out of necessary items and delay the job until new supplies can be obtained. At times, circumstances arise that make it necessary to use more materials than anticipated, so it is better to have materials left over than to run short.

Consider the tools and equipment that you will need for the job and arrange to have them available at the place where the work is to be done, and at the time the work is getting under way. Determine who will use the tools, and make sure the men to whom they are assigned know how to use them properly and safely. Plan to store the materials in an accessible place without creating a safety hazard.

## SUPERVISING WORK TEAMS

After a job has been properly planned, it is necessary to supervise it carefully to ensure that it is completed properly and on time.

Prior to starting a job, make sure your workers know what is to be done. Give instructions clearly, and urge your people to ask questions on any points that are not clear to them. Be sure everyone knows all the pertinent safety precautions and that they wear safety apparel where required. Check all tools and equipment before use to ensure that they are in safe condition. Do not permit the use of dangerously defective tools and equipment; see that they are turned in for repair immediately.

While the job is under way, check periodically to see if the work is progressing satisfactorily. Determine if the proper methods, materials, tools, and equipment are being used. If a man is doing a job wrong, stop him and point out what he is doing wrong. Explain the correct procedure and check to see that he follows it. In checking the work of your men, try to do it in such a way that your men will feel that the purpose of your inspection is to teach, guide, and direct, rather than to criticize and find fault.

The supervisor should make sure that his workers observe all applicable safety precautions and wear safety apparel when required. He should also watch for hazardous conditions, improper use of tools and equipment, and unsafe work practices which could cause accidents and possible injury to personnel or damage to equipment. Many inexperienced workers are heedless of danger, or think a particular regulation is unnecessary. Very often such persons get hurt or cause equipment damage.

When time permits, rotate the men on various jobs. Rotation gives them varied experience and ensures that you will have workers who can fill in if someone is missing from work on a particular day.

A good supervisor should be able to get others to work together in getting the job accomplished. He should maintain an approachable demeanor, and his workers should feel free to seek his advice when in doubt about any phase of the project. Emotional balance is especially important; a supervisor cannot appear panicky or unsure of himself in the face of conflicting forces, nor can he be a person who is pliable with influence. He should use tact and courtesy when dealing with his men and not show partiality to certain members of the work team. He should keep his team informed on matters that affect them personally or concern their work. He should also seek and maintain a high level of morale, keeping in mind that low morale can have a definite effect upon the quantity and quality of work.

The above is only a brief treatment on the subject of supervision. As you advance in your training, you will spend more and more time in the supervision of others. So let us urge that you make a continuing effort to learn more about the subject. Study books on supervision as well as leadership. Also, read journal articles on topics that concern supervisors such as safety, training, and job planning. There is always a need for good supervisors, so consider the role of supervisor a challenge and endeavor to become proficient in all areas of the supervisor's job.

# PREPARING REQUISITIONS

As a crew leader, you should become familiar with requisition forms if they are used in your particular installation. Requisitions are orders from one activity requesting material or service from another activity. Printed requisition forms are designed to provide all the necessary information for the physical transfer of the material and the accounting requirements.

You must list the item's stock number (when available), quantity, and name or description of each item needed. This form is then turned over to your supervisor for his approval and signature, and then forwarded to the purchasing department. The purchasing department obtains the materials requested.

# TIMEKEEPING

In boiler room management, your duties may involve the posting of entries on time cards. Therefore, you should know the types of information called for on time cards and understand the importance of accuracy in labor reporting. Here we are primarily interested in the labor reporting system used in a typical boiler room operation.

In order to measure and record the number of man-hours used during a work day and on various functions, a labor accounting system is necessary. This system must permit the day-to-day accumulation of detailed labor utilization data so that the information required by the person in charge of payroll, and in reports to persons in higher authority can be readily compiled.

Since the system may vary from one installation to another, the suggestions given here can be considered an example.

The boiler room management team must account for all labor expended in carrying out assigned tasks and functions. This accounting must include the work performed by the personnel in the boiler room as well as the work performed by other companies. Labor expenditures must be accumulated under a number of reporting categories. Detailed reporting is required in order to provide management with the necessary data to (1) determine the labor expenditures on project work so that statistical labor costs can be calculated; and (2) compare actual construction performance with estimating standards. It also serves to determine the effectiveness of labor utilization in performing administrative and support functions, both for internal management and for the development of planning standards by higher authority.

For timekeeping and labor reporting purposes, labor is sometimes considered as being in one of two categories: productive and overhead.

## Productive Labor

Productive labor includes all labor that directly or indirectly contributes to the accomplishment of the overall job. Productive labor is accounted for in two categories: direct and indirect labor.

1. Direct labor includes all labor expended directly to the completion of the assigned task.
2. Indirect labor comprises labor that is required for operations, but which does not produce an end product itself.

## Overhead Labor

Overhead labor is not considered to be productive labor in that it does not contribute directly or indirectly to the end product. It includes all labor that must be performed regardless of the assigned job.

During the planning and scheduling of a construction or repair project, each phase of the project considered as direct labor is given an identifying code. Due to the many types of projects that may be encountered and different operations involved, codes for direct labor reporting may vary widely from one activity to another. The crew leader uses direct labor codes in reporting the hours spent by each member of his crew during each work day on assigned tasks.

The crew leader's report is submitted on a daily distribution report form. The report is prepared by the crew leader for each phase of the project that his crew is involved with, and immediately provides a breakdown by man-hours of the activities for the various labor codes for each man in the crew, for any given day, on any project. It should be reviewed at the higher management level and initialed before being sent to the proper level management. It will be tabulated along with all of the daily labor reports received from other departments of the company. It serves as a means by which management can analyze the labor distribution of total manpower resources for any given day, and as feeder information for the preparation of the monthly operations report or any other resource reports required of the unit.

Bear in mind that this information must be accurate and timely, and that each level in the company should review it. The labor report should not serve merely as a feeder report to the boiler room management team; it should also be used by the company to analyze internal construction management and performance.

# SAFETY DUTIES OF A SUPERVISOR

Every supervisor is responsible for the safety of the personnel placed in his charge. Basically, the safety duties of a supervisor revolve around training subordinates, correcting unsafe practices and conditions should they occur, and being prepared to execute certain procedures should a worker be involved in an accident.

## Safety Training

New methods for performing construction and repair operations are being established all the time; therefore, the supervisor must keep himself and his men informed of the latest developments in safety procedures. Moreover, you can never assume that new men on the job are appropriately trained in safety matters. For these reasons, the safety education and training of subordinates is a continuing responsibility of every supervisor.

Every crew leader periodically holds short (approximately 15 minute) safety meetings, during which he briefs his crew on hazards and precautions relating to current work. Although the crew leader is usually responsible for the actual conduct of the meetings, much of the content of the briefing is organized and assembled by a project safety manager.

In addition to these meetings, the crew leader is also concerned with the incidental, day-to-day instruction and training of his men on the job. It is beyond the scope of this text to go into a discussion of teaching or training methods. However, a few words on the crew leader's approach to safety and training at the crew level is appropriate.

The job of achieving safety consciousness in your crew is, like most other supervisory functions, essentially a matter of leadership. In studying and seeking to understand the practical aspects of directing and managing men, many new crew leaders fail to recognize the power of personal example in leading and teaching subordinates. You will soon discover, in this regard, that subordinates are quick to detect any difference between what you say and what you do. You cannot reasonably expect your men to measure up to standards of safety conduct and awareness if you do not constantly demonstrate them. It is not enough to be knowledgeable in the various aspects of construction and maintenance safety. As a supervisor, you must make your genuine concern for the crew members visible at all times. Leadership by example may not be the only technique of leadership, but it is one of the most practical and time-proven methods of management available to you.

## Accident Reporting Procedures

A well planned and properly executed crew safety program will prevent accidents. Nevertheless, you must be prepared to carry out certain procedures should a man in your charge be injured or otherwise involved in an accident. It might be mentioned that for the purpose of accident reporting, an accident is defined as "Any unplanned act or event which results in damage to property, material or equipment, or personnel injury or death." For purposes of this discussion, an injury may be defined as "Any physical impairment which prevents a person from performing his regularly established duty or work for a period of 24 hours or more on the day of the injury." If an injury requires first aid or medical attention and the man can be returned to duty within 24 hours, the formal investigation and reporting may not be necessary, depending upon the requirements at your facility. This is not to say, however, that *you* need not investigate these kinds of accidents on your own initiative. Small or seemingly harmless accidents often repeat themselves with more serious results. If an improper practice or condition exists, you must obviously identify and correct it before a serious accident occurs.

When one of your men is injured, the first priority is to arrange for prompt medical treatment. At the same time, you must take steps to prevent additional injuries or damage by shutting off power, stopping equipment or machinery, posting guards, and so on. Next, see that the boiler room management is notified of the accident. If the accident is one that causes a physical impairment that will prevent the man from performing his regularly established duties for a period of 24 hours or more, then it is your responsibility as crew leader to investigate the circumstances and submit a written report.

# OPERATIONAL RESPONSIBILITIES OF A CREW LEADER

As crew leader you will probably be the on-the-job supervisor in charge of such operations as installation and start-up, feed water systems, testing, condensate system protection, idle boiler care and lay-up, steam boiler blowoff and blowdown systems, daily boiler log, and steam quality. The equipment manufacturer's recommendations should be followed for all of these procedures. One of these operations, that of idle boiler care and lay-up, is presented below as an example of your responsibility during such procedures.

The following section is reproduced with the permission of the American Boiler Manufacturers Association, Arlington, Virginia.

# IDLE BOILER CARE AND LAY-UP

Corrosion damage to commercial boilers is most often the result of improper lay-up during nonoperating periods. Substantial damage can occur in only a few days if proper precautions are not taken. This damage is irreversible and will reduce boiler reliability, increase maintenance costs, and eventually shorten the useful life of the boiler. Both the water/steam sides and the fire/gas sides must be protected.

Idle boilers are very vulnerable to attack when air contacts untreated wet metal surfaces. To prevent corrosion, therefore, the boiler metal must be protected by either keeping the surfaces completely dry or excluding all air from the boiler.

In addition to the corrosion damage that occurs, the metal particles that are released will form an insulation scale in an operating boiler when the system is returned to service. These corrosion products will deposit on critical heat transfer areas of the boiler, increasing the potential for localized corrosion and overheating during system operation.

Proper lay-up techniques must be used for idle equipment even if it has never been in operation. Before preoperational lay-up, the boiler should be cleaned with an alkaline detergent solution to remove preservatives, oil, and grease. Use the recommended procedures for cleaning.

## Taking Boilers Off-Line

In operation, boiler water contains suspended solids (or mud) which are held in suspension by water circulation and the action of the treatment chemicals. Unless care is exercised when draining, these suspended solids will settle out on the boiler surfaces and air dry to an adherent deposit, sometimes requiring turbining of the boiler. In addition, unless the deposits are examined carefully, it may be incorrectly assumed that the deposits are scale which formed during operation. Therefore, in order to judge the effectiveness of the water treatment program and to eliminate unnecessary boiler cleaning, proper care is imperative during shutdown.

**Preshutdown Precautions.**    For a period of three to seven days before shutdown, the manual blowdown should be increased. During this period, the lower conductivity or chloride limit should be observed as a maximum. The feed of internal water treatment must be increased to maintain specific residuals. Continuous blowdown should be kept to a minimum so the reduction in solids is achieved by increased manual blowdown.

If it is necessary to use a draw-and-fill method for cooling, the pressure should be lowered, and cooling should be at the rate recommended by the boiler manufacturer. Care should be taken to maintain recommended treatment balances during this process. The feed water should be deaerated.

**Wash Down.**    As the boiler is being drained, the manhole(s) and handholes should be knocked in and a high-pressure hose employed to wash out sludge. With this procedure, the sludge is removed while still in a fluid form.

If the boiler cannot be washed immediately on draining, the remaining heat in the boiler setting may cause baking of the residual sludge. The boiler should not be drained until cool. However, never leave the boiler filled with water for any extended period of time without taking measures to prevent corrosion and pitting.

## Steam Boilers

There are two basic procedures for laying up a steam boiler: wet and dry. The choice between the wet and dry methods of lay-up depends upon several factors:

1. The possibility that the boiler may need to be placed in operation on short notice
2. The disposal of lay-up solutions
3. The possibility of it freezing

Wet lay-up is recommended for short outages (30 days or less) and has the advantage of allowing the boiler to be returned to service on short notice. But, this method can be a problem if the boiler will be exposed to freezing conditions.

Dry lay-up is recommended for long idle periods (greater than 30 days), but is practical only if the boiler can be drained hot (120 to 170°F) or if external drying can be provided.

**Short Term (30 days), Wet Lay-Up Steam Boilers.**   In the wet lay-up procedure, the boiler is completely filled with chemically treated water and sealed to prevent air in-leakage. Nitrogen gas under slight pressure can also be used to displace air and "blanket" the boiler surfaces from corrosion.

The steps below should be used to wet lay-up a boiler.

*Step 1A.*   At least 30 minutes before the boiler comes off the line, add the following chemicals:

Sodium Sulfite—30 lbs/gal water
A Polymeric Sludge Dispersant—1 lb/1,000 gal water
Caustic Soda—3 lbs/1,000 gals water

*Step 1B.*   If the boiler has been out of service for cleaning or has never been on-line, use the following procedures:

1.  Select the highest quality water available to lay-up the boiler. Steam condensate, softened water, filtered fresh water, and boiler feed water are generally considered acceptable for lay-up. Raw city water should not be used.
2.  Prepare the chemical solution in a separate tank using the concentrations listed in Step 1A. Use proper safety procedures. Add the concentrated lay-up solution to the boiler during the time it is being filled. The solution can be added using either the operational or the chemical feed line.
3.  After the boiler is filled and the lay-up solution has been added, operate the boiler at low fire for at least 30 minutes to obtain circulation and mixing of chemicals.

*Step 2.*   After filling, the boiler must be tightly blanked or closed. Electric energy to the unit must be cut off. All vent valves are operated as needed so that the boiler can be completely filled with the required solution. To prevent in-leakage of air, pressurize with 5 psig nitogen through a suitable connection during the lay-up period. Follow all safety procedures. An alternative is to install a 55-gallon drum or auxiliary vessel, fitted with a cover and containing properly treated water, on top of the steam drum. The drum or vessel should be connected to an available opening such as a vent line at the top of the boiler to create a hydrostatic head. The tank will provide a ready visual check of water level loss or in-leakage during lay-up. (See Figure 14–1.)

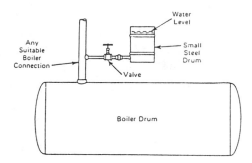

**Figure 14-1.**
Arrangement for keeping boiler completely full of water during wet lay-up (Courtesy of The American Boiler Manufacturers Association)

*Step 3.* The water in a laid up boiler must be tested weekly to make sure that the proper levels of sulfite and alkalinity are being maintained. To do this, take a sample of the boiler water from the surface blowdown line or some other high point. The test results should be:

200 ppm minimum, sodium sulfite (as $SO_3$)

400 ppm minimum, phenolphthalein alkalinity (as $CaCO_3$)

If the tests show that the chemical concentrations have decreased below the recommended minimum, a practical procedure is to add additional lay-up solution to the drum and inject it into the boiler by lowering the boiler water level. To mix the chemical, operate the boiler at low fire, and then follow the procedures previously described under Step 2 for complete filling of the boiler. Attention should be directed to valve maintenance so that untreated water does not accidentally enter the boiler and cause dilution of the properly treated lay-up solution.

(NOTE: An alternate wet lay-up procedure is to pipe clean continuous blowdown water from a properly treated operating boiler into any convenient bottom connection on the idle boiler, allowing excess water to overflow from the vents and into the sewer. This method will insure a continuous complete fill with warm, properly treated water and will prevent air in-leakage by keeping the water side slightly pressurized. In addition, it may provide enough heat to keep the fireside dry, and also provide freeze protection.)

**Long Term (over 30 Days), Dry Lay-Up, Steam Boilers.** In storing a dry boiler, trays of moisture-absorbing chemicals are placed in the boiler, then all openings are sealed. In an alternate dry procedure, the boiler is drained, dried as completely as possible, the manholes closed and all connections tightly blanked or closed, and then the boiler pressurized with nitrogen to prevent air from getting inside. The success of this procedure depends on a thorough drying of all boiler surfaces after draining and the exclusion of air during lay-up by pressurizing the boiler with nitrogen. Use the proper safety precautions.

*Caution:* Boilers laid up dry shall be tagged with information to indicate that the unit must not be operated until the moisture-absorbing materials are removed and the boiler refilled.

The following steps are to be taken for dry lay-up:

### *Desiccant Method*

1. After the boiler has cooled, drain it completely.

2. Flush the boiler thoroughly and inspect it. A boiler containing porous moisture-retaining deposits should not be laid up dry because of the possibility of under-deposit corrosion occurring. If cleaning is needed, this must be done before the boiler is laid up.

3. Dry thoroughly. Circulated warm air should be used in drying the metal.

4. Put quick lime (not dehydrated lime) or a commercial grade silica gel on wooden or plastic trays and place them inside the boiler drums or shell. The trays should be placed so that air may circulate underneath them. Use at least 5 lbs of quick lime per 1,000 lbs/hr steaming capacity or 5 lbs per 30 boiler horsepower. If silica gel is used, the quantities recommended are 8 lbs of silica gel per 1,000 lbs/hr steam capacity or 8 lbs per boiler horsepower. Do not fill the trays more than half-full.

5. Close all manholes and blank or close all connections on the boiler as completely as practicable to prevent in-leakage of air. An option is to pressurize the boiler with 5 psig nitrogen after the desiccant is installed.

6. Inspect the water side of the boiler every three months for evidence of active corrosion. Check the desiccant and replace it if necessary. If the boiler has been pressurized with nitrogen, it should *not* be entered until the safety precautions described below for nitrogen have been taken. Do this work as quickly as possible to minimize the entry of humid air into the boiler.

7. If the desiccant is wet, dry out the boiler again before replacing the desiccant and repressurize with nitrogen.

*Caution:* Boilers laid up dry shall be tagged with information to indicate that the unit must not be operated until the desiccant is removed and the unit refilled.

### Nitrogen Blanking Method

1. (A) Drain the boiler, but before the pressure falls to zero pressurize with 5 psig nitrogen through a suitable connection while draining. Maintain this nitrogen pressure throughout draining and subsequent storage.

   (B) An alternate method is to completely dry a clean boiler and then purge air from the boiler and pressurize with 5 psig nitrogen. Be aware that all metal surfaces not completely dried are vulnerable to corrosion, particularly if oxygen is present. Drying can be aided by employing low auxiliary heat or blowing hot, dry air through the boiler.

2. If a boiler has been down for repairs and is to be laid up with nitrogen, it should be operated to repressurize the system with steam and then drained and pressurized with nitrogen as previously discussed in 1(A), or follow the procedure discussed in 1(B).

3. All connections and valves must be blanked or tightly closed.

   (NOTE: Operating boilers must be removed from service properly to minimize adherence of boiler water suspended solids on boiler metal surfaces. This can be accomplished by immediate flushing with hot, pressurized water, while the waterside surfaces are still wet. If a boiler contains deposits that formed during operation or due to improper shutdown procedures, mechanical or chemical cleaning is required.)

**Safety Precaution.**    Using nitrogen for blanketing is recommended in many of the lay-up procedures. Nitrogen will not support life; therefore, it is essential that proper precautions be taken before such equipment is entered for inspection or other purposes. These precautions shall include disconnecting of the nitrogen supply line, thorough purging and venting of the equipment with air, and testing for oxygen levels inside the equipment. Appropriate caution signs shall be posted around the equipment to alert personnel that nitrogen blanketing is in use.

**Returning Idle Steam Boilers to Service.**    Instructions for returning steam boilers to operation after lay-up are listed below.

*After Wet Lay-Up.*    To start-up an idle boiler after wet storage use the following procedure:

1. If the boiler was pressurized with nitrogen, disconnect the nitrogen supply sources and vent the boiler.
2. Using the blowdown valve, partially drain the boiler water and make-up with feed water to dilute any chemical residuals to the proper operating levels.
3. After the boiler water concentration and water level is returned to the proper operating conditions, the boiler can be started up in the normal manner.

*After Dry Lay-Up.* To start-up a boiler after dry lay-up, use the following procedure:

1. If the boiler was pressurized with nitrogen, disconnect the nitrogen supply. Vent the nitrogen in a safe manner external to the building and away from fresh air intakes. Then thoroughly purge the boiler of nitrogen with dry air. This is mandatory before personnel can enter the equipment since nitrogen (or lack of oxygen) will not support life.
2. Remove any and all desiccant from the boiler.
3. Follow the manufacturer's start-up procedures.

## Lay-Up of a Hot Water Boiler

Prior to start-up, a new boiler which contains oil, grease, cutting solutions and other contaminants, should be cleaned. Likewise, a hot water boiler containing very dirty water or an untreated system should be flushed and cleaned prior to lay-up. Follow the recommended procedures for cleaning the boiler.

There are two basic procedures for laying up a hot water boiler: wet lay-up, and dry lay-up. The choice between the wet and dry methods of lay-up depend on:

1. The possibility that the boiler will soon be returned to service
2. The possibility of freezing

Wet lay-up is normally recommended for hot water boilers and their associated systems. But this method can be a problem if the boiler will be exposed to freezing conditions.

Dry lay-up is only suggested for indefinite idle periods (frequently over

1 year), but is practical only if the boiler and system can be drained hot (over 120°F).

**Wet Lay-Up.**   The boiler and system are completely filled with chemically treated water and sealed to prevent air in-leakage. The following is the recommended procedure for wet lay-up:

1. The boiler and system should be completely filled with good quality water.
2. In a separate tank, prepare either of the following chemical solutions:
   (A) Borax          8 lb/1,000 gal of holding capacity
       Nitrate        6 lb/1,000 gal of holding capacity
   (B) Sodium Sulfite      5 lb/1,000 gal of holding capacity
       Sodium Hydroxide   3.5 lb/1,000 gal of holding capacity
   The solution for alternative (A) may be the same chemical program on which the boiler is maintained while in service.
3. Fill the boiler with the chemical solution, providing for adequate mixing. The boiler and system should be completely filled, eliminating all air pockets.
4. After filling, the boiler must be tightly blanked. All vent valves are operated as needed so that the boiler can be completely filled with the required solution.
5. The water in a layed up boiler must be tested monthly to make sure that the proper chemical levels are maintained and that the system is completely filled with water. To do this, take a sample of the boiler water from the surface of the boiler. The test results should be:
   (A) 250 ppm minimum, nitrate (as $NO_2$)
       pH between 9.5 and 10.5
   (B) 200 ppm minimum, sodium sulfite (as $SOP_3$)
       400 ppm minimum, phenolphthalein alkalinity (as $CaCO_3$)
   Add additional chemicals as needed, insuring that they are mixed properly by firing the boiler lightly to circulate the water. Be certain that no air is introduced into the system when the chemicals are added.

**Long Term Dry Lay-Up of the Hot Water Boiler.**   The long term dry lay-up of a hot water boiler is identical to the dry lay-up of a steam boiler. This method should be used only when the boiler is not expected to return to service. See previous instructions.

  *Caution:* A boiler that is laid up dry should be properly tagged and identified that the boiler is dry and desiccants are in place.

**Returning Idle Hot Water Boiler to Service.**    The procedures for returning a laid up hot water boiler to service are describe below.

*After Wet Lay-Up.*    To start-up an idle boiler after wet storage, use the following procedure:

1. Isolate and partially drain the boiler. Make-up with fresh water to dilute chemical residuals to operating levels.
2. After the concentration is returned to the proper operating conditions, the boiler can be started up in the normal manner.

NOTE: Some sanitary districts restrict the amount of nitrate which may be placed in the sewer system. Check the applicable regulations prior to disposing of the chemicals.)

*After Dry Lay-Up.*    To start-up a boiler after dry storage, use the following procedure:

1. If the boiler was pressurized with nitrogen, disconnect the nitrogen supply, and vent the nitrogen in a safe manner, external to the building and away from fresh air intakes. Then thoroughly purge the boiler of nitrogen with dry air. This is mandatory before personnel can enter the equipment, because nitrogen will not support life.
2. Remove any and all desiccant from the boiler.
3. Follow the manufacturer's recommended start-up procedures.

**Safety Precautions.**    Safety is a critical factor that is often overlooked. The following safety factors must be observed during a boiler lay-up:

1. Obtain Material Safety Sheets from the supplier for all chemicals used for cleaning and or lay-up.
2. Read and follow the recommended handling procedures and precautions.

# Fireside Lay-Up

Fireside lay-up procedures are designed to keep metal surfaces dry. Moisture and oxygen cause corrosion by forming acids with any fuel deposits left in the boiler. These acids attack steel. Precautions taken during lay-up will inhibit metal degradation and prolong the boiler life.

The deposits themselves can cause three types of problems. First, as deposits, they can produce corrosion at the point where the metal and the deposit meet. Second, the deposit may trap fly ash which adds to the bulk of the deposit. Third, fly ash constituents such as iron, vanadium, and sodium may react with sulfur compounds to form highly corrosive, low pH deposits.

All of these problems can be prevented by a good fireside maintenance program. Fireside cleanup for coal- and oil-fired boilers may be done by water washing the tube surfaces with lances or high-pressure hoses. Tenacious deposits may have to be sandblasted from the tube surface. For boilers burning natural gas, brushing the tubes and vacuuming up the debris is probably all that is necessary.

But cleaning the deposits from the tubes is not all it takes. Clean tube surfaces are also vulnerable during lay-up to rust and corrosion. This is because the water used to clean the fireside reacts with sulfur compounds in the ash deposits to form sulfuric and sulfurous acid. Iron or vanadium, and fuel impurities will speed up this reaction and intensify the corrosion.

To prevent these problems, two methods of lay-up are used: hot or cold. Cold lay-up is better for extended outages (longer than three months) or if boiler repairs are needed. For minor repairs, a short outage, or when keeping the boiler idle, a hot lay-up is preferred.

With either method of lay-up, if significant deposits are present, the fireside should be cleaned. If water washing is undesirable, an oil dispersible magnesium-based additive can be added continuously for two weeks prior to shutdown to neutralize the corrosiveness of the deposits. During lay-up the boiler should be inspected monthly to check for trouble spots and active corrosion sites.

**Hot Lay-Up—Boiler Idle.**    For hot lay-up, metal surfaces should be kept at 170°F or higher to prevent moisture entering the system. The temperature can be controlled by using an auxiliary heat source.

Natural gas or electric air heaters can be used to maintain boiler temperatures at 170°F. Circulation of hot blowdown water through the boiler waterside may be used if enough blowdown water is available to keep the fireside surfaces dry.

**Cold Lay-Up.**    The potential for corrosion is much greater with cold lay-up because of the lower temperatures. A metal surface below 300°F with 10 ppm $SO_3$ present will be cool enough to cause condensation of sulfuric acid, opening the door to corrosion. During cold lay-up the metal temperatures are usually far below 300°F, so the conditions are perfect for sulfuric acid formation.

To prevent this acidic corrosion, wash down the boiler fireside surfaces with water when the unit is off-line and the water temperature has dropped below 140°F. A five percent solution of an alkaline chemical like soda ash can be used to neutralize acidic deposits. Washing removes the ash and impurities which can contribute to corrosion. *Drain all wash water from the boiler.*

Metal surfaces should be kept dry by using heat lamps, desiccants, or dry, warm air circulation. Seal the furnace to prevent moist air or rain from entering.

# REVIEW QUESTIONS

1. What is the first thing you should do when it is determined that a specific job is to be done?
2. As a boiler room manager, what must you know about the personnel under you?
3. Should several jobs be planned in advance?
4. What type of goals should you set for the crew?
5. Should the same rate of production be expected both during an emergency and during normal work conditions?
6. What precaution should be taken when estimating materials?
7. What should you do when men are working with tools?
8. What should the boiler room manager strive to do when checking the work of his crew?
9. What should the crew leader watch for?
10. What type of demeanor should a supervisor have?
11. What should a good supervisor strive to maintain among his workers?
12. How is material for the boiler room usually ordered?
13. What are the two types of labor?
14. In regards to safety, what are the duties of a supervisor?
15. What method of leadership is the most practical and time-proven?
16. Define a personal injury.

17. What should be done to the blowdown frequency beginning three to seven days before shutting down a boiler?

18. Name two procedures used for laying up a steam boiler.

19. Why must the water in a laid-up boiler be tested weekly?

20. What should be done to a new boiler prior to start-up?

# CHAPTER 15

# BASIC WATER SYSTEMS

## Learning Objectives

The objectives of this chapter are to:

- Describe different classifications of hot water systems
- Introduce the reader to the different types of accessories used in hot water boiler systems
- Show operation of the different types of hot water boiler system controls
- Describe the uses of the different hot water boiler system components.

Systems in which water is heated at a central plant and circulated through pipes to radiators, convectors, or unit heaters are called hot water heating systems. These systems are very useful carriers of heat and may be classified according to their operating temperatures, such as low, medium, or high temperature. In this chapter we will cover the low-medium temperature range and for practical purposes will further classify the systems into gravity or forced. The purpose and function of components will be discussed, as well as the advantages and disadvantages of the different types of hot water systems.

The water temperature for low-temperature hot water (LTHW) heating systems ranges from 100°F to 220°F.

For outside distribution, large space heaters, absorption refrigeration equipment, and industrial purposes, the design supply will normally be of the medium temperature range from 220°F to 300°F.

The descriptions of the following classes and types of systems will pertain primarily to low and medium temperature hot water.

# CLASSIFICATION OF SYSTEMS

There are two general types of low-temperature hot water heating systems. The first type is the gravity feed system in which water circulation depends upon the weight difference between the hot column of water leading to the point of use, and the relatively cooler, heavier column of water returning from the point of use. The second type is the forced-circulation system in which the water is circulated by a power-driven pump.

Hot water systems are also classified according to their temperature operating range. Low-temperature hot water (LTHW) heating systems have a temperature range of 100°F to 220°F. Medium temperature ranges from 220°F to 300°F. For simplicity, we will classify the systems as either gravity or forced circulation.

## Gravity Circulation

In gravity systems, the flow of water is caused by the difference in the weights of the column of hot water in the supply risers and the column of cooler water in the return risers of the system. The density (weight) of water decreases as its temperature rises; therefore, the size of the head available for circulation depends on the temperature difference of the water in the supply and return risers. There are several different types of gravity flow systems which we will discuss.

## Forced Circulation

In forced-circulation hot water systems, the flow of water is caused by a water circulating pump. These pumps are generally of the centrifugal type. Their only purpose is to move the water through the system; they do not maintain the pressure inside the system. The circulating pump is generally located on the return water main line close to the boiler inlet; however, they are sometimes installed in the supply main at the boiler outlet.

## One-Pipe System

In this type of system, one single main line is carried entirely around the building. The water flows from the boiler outlet and back to the boiler inlet. Some of the water is directed to the point of use by a bypass to the individual units. (See Figure 15-1.) Thus, the size of the main line remains constant from the boiler outlet to the boiler inlet. The supply risers are taken off the main line to feed the heat-transmitting units and the returns from the units are reconnected to the main line, a short distance along the line.

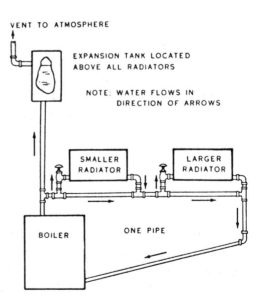

**FIGURE 15-1.**
A one-pipe circulation system

This can be done without interfering with normal operation because the differential pressures of the heating unit inlets and outlets cause the water to flow. This differential pressure is produced by the constant drop, from the pump discharge to the pump suction, of the pressure head in the main line. Therefore, if a supply riser is taken from any point and the return is reconnected into the main at the proper distance along the line, there will be enough pressure difference between the two points to produce water circulation through radiation. Usually, 8 or 10 feet between these connections will produce the desired circulation. If special flow fittings are installed at the inlet of the supply riser or at both the supply and return branches, connections can be placed within one foot of each other. Some designers increase the size of the radiation to compensate for the gradual temperature drop in the main line.

## Two-Pipe System

The two-pipe system uses two main lines: The supply main feeds water to the risers that serve the heating units, and the return main line collects the water which is returned from those units. The two mains run side-by-side; the supply main decreases in size and the return main increases in size at each branch connection. Since the heating units of a two-pipe system are connected in parallel, it requires a minimum pumping head. Also, if throttle valves or restricting orifices are used in the risers, the flow of water through the individual units can be adjusted easily over a wide range. However, the two-pipe system requires more pipe and pipe fittings than a one-pipe system. Two-pipe systems are classified as direct-return and reverse-return.

**Direct-Return System.**   The heat-using devices of a two-pipe, direct-return system are connected in parallel. Nevertheless, the water taken from the main to feed the first radiation device is returned first; that removed for the second radiation device is returned second; and so on successively throughout the complete system. Since this procedure causes a progressively greater frictional loss in each additional circuit, the flow circuits become hydraulically unbalanced. This condition may cause the first radiation device to have a greater flow than is required to develop its full capacity, while in a large system, the flow through the last device may be so small that practically no heat is delivered. To balance the system, carefully select pipe sizes to compensate for the differences in the length and the consequent friction loss of each circuit. (A balanced circuit must have the same pressure drop through each piping circuit at the designed flow rate.) Restricting orifices or throttle valves can be installed to correct the flow distribution and to balance the system after it is placed in service. (See Figure 15–2.)

**Reversed-Return System.**  In the two-pipe, reversed-return system the water taken from the main line to feed the first radiation device is returned last to the main line; the water supplied to the last radiation device is returned first. As a result, all unit circuits are of approximately the same length, a condition which is conducive to system balance. The reversed-return system may require more pipe than the direct-return system. However, its inherently better flow distribution and natural balance, without the aid of additional valves or orifices, compensate for the additional cost. (See Figure 15–3.)

## Open-Tank Systems

Every hot water heating system should have an expansion vessel to handle the expansion and contraction of the water as it is heated and cooled.

In open-tank systems, the expansion tank is located at the highest point in the system and freely vented to the atmosphere. Normally, these systems are limited to installations with operating temperatures of 180°F or less.

**One-Pipe, Open-Tank System.**  In this type of system, a one-pipe loop, that is a single pipe, leaves the boiler and returns to it. The heat-using devices

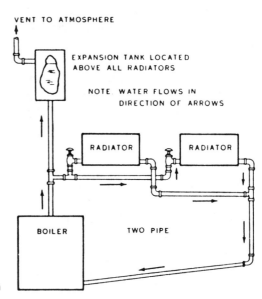

**FIGURE 15-2.**
A two-pipe direct-return system

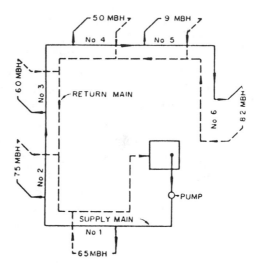

**FIGURE 15-3.**
A two-pipe reversed-return system

each have one connection from the loop and another leading back to it. (See Figure 15–4.)

This type of system is relatively simple to install and is usually moderate in cost. Proper operation of the system depends upon the design of the system allowing only for a small temperature drop through the heat-radiating devices, so that the water reaching the last device is not too much cooler than the water reaching the first device.

**Two-Pipe, Open-Tank System.**   In this type of system, the inlet side of the radiation device is connected to one pipe, and the outlet side is connected

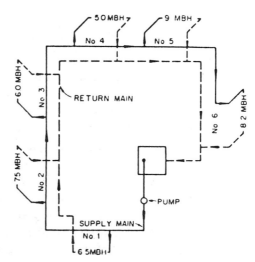

**FIGURE 15-4.**
A diagram of a one-pipe, open-tank, gravity hot water distribution system

to a returning line. The majority of hot water gravity systems are of the two-pipe, open-tank type.

The problem of balancing water flow in two-pipe, gravity systems, as well as forced systems, can be greatly reduced by use of a reversed return. In the reversed-return system, the water taken from the supply main to feed the first radiation device is returned last to the return main line; the water supplied to the last device is returned first. As a result, the water travels in circuits of approximately the same length, a condition which is conducive to good system balance. This is in contrast with the direct-return system, whereas the water taken from the main to feed the first radiation device is returned first; that removed for the second radiation device is returned second, and so on, successively throughout the complete system.

Where a boiler is installed on the same level as the lowest radiation device, a downfeed system is used. The supply riser is carried as high as practical, thus simplifying the piping to the radiation devices on the same level with the boiler, and increasing the heat available to provide water circulation.

## Closed-Tank Systems

The expansion tank for the closed system is a pneumatic compression tank, which is not open to the atmosphere. Therefore, hot water heating systems using it can be pressurized and operated over a wide range of temperatures and pressures. Thus, when the saturation pressure is raised to correspond to the desired temperature, the system can be operated above 212°F.

In a closed expansion tank, as the water temperature lowers, the water contracts, air in the tank expands, the excess water returns to the system, and the pressure drops. The tank must be large enough to keep a reservoir of compressed air above the water level to cushion the excess water that enters. Thus, the tank must provide space for change in both water and air volumes.

If the tank is too small or contains insufficient air, the water will expand and the system pressure may increase above the permissible level. This will cause the relief valve to open and waste water. Also, as the temperature drops, the water contracts and the pressure may drop below the permissible level. Air will not vent from the system and additional air may be drawn in, if the high points of the system have automatic air vent valves.

The capacity of the tank depends basically on the amount of water in the system and the operating temperature range. A minimum volume of six percent of the total system water (boiler, piping, and other equipment) is recommended for open tanks. The size of closed tanks is usually calculated by formulas which take into account the volume of water in the system, tank operating pressure, temperature operating range, pressure in the tank when water first enters (usually atmospheric), and the initial fill (or minimum) pressure in the tank. It is recommended that the ASME (American Society of Mechanical Engineers), "Low-Pressure Heating Boiler Code," be referred to for minimum capacities of expansion tanks in closed systems.

The location of closed tanks depends on the type, size, and the design of the installation. Regardless of location, the point of junction between the tank and the system is under constant pressure, even when the circulating pump is not operating. When possible, the expansion tank should be connected by a direct pipe to the highest point of the boiler. This arrangement will let air pass easily to the tank. (See Figure 15–5.) In closed tanks, a water gauge and air inlet, water inlet, drain and relief valves permit operators to observe and adjust the proportion of air in the tank. In open tanks overflow and vent pipes are used instead of relief and air inlet valves.

**One-Pipe, Closed-Tank Gravity System.**    The one-pipe, closed-tank, gravity type of hot water heating and distribution system is similar to the one-pipe, open-tank gravity system, except that the expansion tank for the closed-type system is a pneumatic compression tank, which is not open to the atmosphere.

The tees, by an injecting action on the supply and ejecting action on the return, increase water circulation through the heating devices. These tees also aid stratification of the hot and cold water within the main line. They

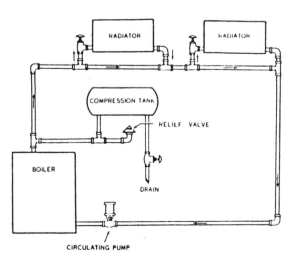

**FIGURE 15-5.**
A diagram of a one-pipe, closed-tank, forced-circulation hot water heating system

are designed to take the hottest water from the top of the main line and to deposit the colder water on the bottom of the line.

**Two-Pipe, Closed-Tank System.**   The general arrangement of the piping in these types of systems is the same as for the two-pipe gravity systems. The relative locations of the expansion tank, circulating pump, and flow control valve are essentially in the same location. (See Figure 15-6.)

In a closed system, makeup requirements are low. Normally, the feed water system of a hot water installation will consist of reducing valves, relief valves, manual feed connections, check valves, and circulators.

## SYSTEM ACCESSORIES

The following is a description of the accessories that are used on hot water distribution systems.

## Open Expansion Tank and Closed Compression Tank

These tanks are one of the most important components to a hot water heating system. They have been covered thoroughly in the preceding section of this chapter.

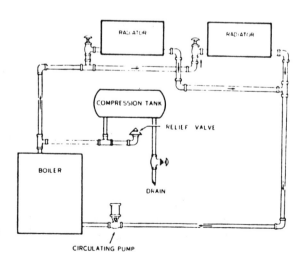

**FIGURE 15-6.**
Diagram of a two-pipe, closed-tank, forced-circulation hot water distribution system

## Pressure-Reducing Valves

A pressure-reducing valve is generally installed in the water makeup or cold water line to the boiler. This valve automatically keeps the closed system supplied with water at a predetermined safe pressure. The maximum permissible pressure for a standard hot water heating system is 30 psig; therefore, the reducing valve setting should be as low as possible. Valves are usually factory set at 12 psig pressure; this equals a static head of 276 feet of water (suitable for buildings with 1, 2, or 3 stories). However, if the static head of the system is high, boilers with a higher operating pressure may be required, and the reducing valve is set correspondingly higher.

## Pressure Relief Valves

All hot water heating systems must be provided with a pressure relief valve. Otherwise, water expansion can subject the system to excessive pressures if connections to the expansion tank close because of freezing, or for other reasons; if the tank becomes completely filled with water (a noncompressible fluid); or if the air volume in the tank is inadequate for the necessary expansion.

Each system has a conventional hot water, pressure relief valve. The valve has a spring-loaded diaphragm that raises the valve seat when the water pressure exceeds the predetermined maximum safe level (usually 30 psig).

## Flow Control Valve

This is a special type of valve which is used to prevent the gravitational flow of water through the system when the circulating pump is shut off. It is located on the supply main line just above the boiler outlet connection. This valve operates under the same principle as the check valve, only it prevents gravitational flow of the water instead of reverse flow.

## Check Valves

The check valve automatically operates to prevent a reversed flow of water into the supply line. Two types of check valves are used. (See Figure 15-7.) These are known as the lift-check and the swing-check valves. In both,

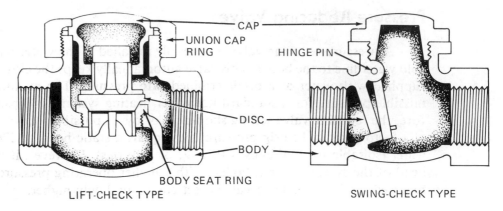

**FIGURE 15-7.**
Check valves

the fluid enters below the seat and lifts the disc, permitting the fluid to flow through the system. As soon as the flow stops, the disc is closed against the seat by its own weight. If the fluid tends to reverse direction, the valve automatically shuts tightly, thereby preventing a flow through it.

Special flow fittings, special tees, are available for use in one-pipe systems to deflect main line water into the radiation devices or the branch circuits. These tees impose the necessary pressure drop in the main line between the supply and are connected to deflect the quantity of water required by the radiation equipment and the system temperature.

## Gate Valves

Gate valves are placed in lines where an unrestricted flow is important. They are not suited for flow regulation or throttling service because when so used the gate may vibrate and chatter, damaging the seating surfaces. Also, throttling will cause erosion of the lower edge of the seat rings. Gate valves are built in four different types according to the difference in their stem or gate construction. Stems are either rising or nonrising; gates are double-disc or wedge-disc.

## Manual Feed Connection

Usually, a connection for manual feed is made in the return main line near the boiler inlet. The feed is regulated by a manually operated globe valve.

Where it may become necessary to throttle the flow, a globe-type valve is used. The flow passing through the body of this valve changes direction, thereby creating greater frictional resistance. Globe valves generally have inside stems and rising screws, and plug or disc-type valves and seats. Discs are metal or composition. Metal discs are suited for throttling service; composition seats are not, because of the severe cutting action on the disc face. These valves are available with renewable metal seats and discs. (See Figure 15–8.)

## Plug Cocks

The primary function of the plug cock is to regulate the flow rate to control the system temperature. (See Figure 15–9.) Plug cocks may be used in lines that are not subject to excessive heat, but they should be lubricated and tested frequently to insure leak-free operation. These cocks are usually found on the return line of each branch to balance the temperature within that circuit.

## Pipe Anchors, Hangers, and Supports

Pipe anchors, hangers, and supports carry the weight of the pipe, valves, fittings, and fluid in a piping system. In combination with expansion joints and bends, they control and guide pipe expansion. They must be strong

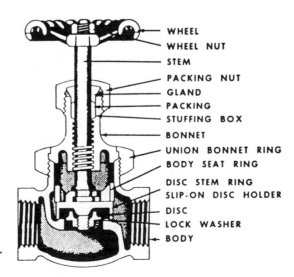

**FIGURE 15-8.**
Globe valve

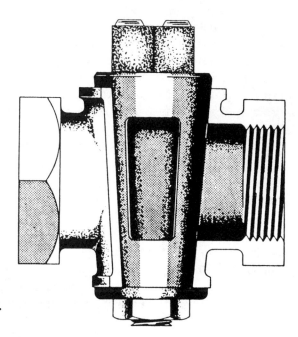

**FIGURE 15-9.**
Plug cock

enough to support the full, unbalanced pressure of the fluid and expansion strains, as well as the weights of piping and fluid. In addition, they must prevent or minimize excessive vibrations and shock.

**Anchors.**    Pipe can be anchored in an underground trench by embedding the ends of a structural steel shape in the trench walls and resting the pipe on it. The pipe is attached to the steel shape by a heavy clamp with bolts which is welded to the pipe. If the pipe is overhead on poles, use wire rope and turnbuckles to guy the ends of the structural shape in both directions to concrete deadmen. The best location for anchors is midway between expansion loops and joints.

**Hangers and Supports.**    The open markets offer a variety of hangers and supports. They are usually made of steel and include pipe rings, ring clamps, beam clamps, brackets, U-bolts, saddle supports, pipe rools, and spring hangers. The spacing between the hangers is important. With too much space, the line will sag, permitting condensate to collect (in steam lines) and reduce system efficiency. Also, be sure that the line is pitched in the direction of the flow and that enough hangers and supports are used to minimize vibration.

# Expansion Joints

Expansion joints and loops in long heating lines are convenient devices for handling pipe elongation caused by the expansion. There are five principle types: (1) slip joints, (2) bellows joint, (3) swing joint, (4) expansion loops, and (5) ball joints.

**Slip Joints.** In this type of joint, a female member is slid over a male member while the joint is held tight by packing and permits expansion and contraction. (See Figure 15–10.) The type of packing determines the temperature to which the joint may be subjected.

**Bellows Joint.** The bellows joint has a metal bellows which flexes as expansion and contraction occurs. (See Figure 15–11.) The joint consists of a thin-walled, corrugated copper or stainless steel tube which is clamped between the flanges. Rings help keep the corrugations under relatively high pressure. The pipes should be supported and guided to keep misalignment to a minimum.

**Swing Joint.** The swing or swivel joint is most often used to allow for expansion. It is better adapted to screwed joints. When it is used with welded elbows, the swing introduces torsional strains in the elbows and the swing piece.

**Expansion Loops.** The expansion loops absorb the expansion by introducing U or Z loops in the pipe line.

**Ball Joints.** Ball joints are often used instead of expansion loops because they require less space and material. Ball joints have four basic parts:

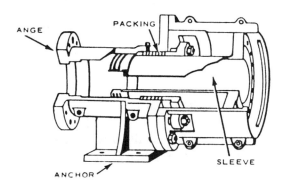

**FIGURE 15-10.**
Slip joint

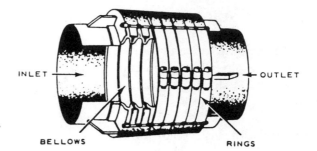

**FIGURE 15-11.**
Bellows joint

1. The casing, or body, to hold gaskets and ball
2. The ball (a hollow fitting shaped externally like a ball at one end, which is inside the casing, and threaded, flanged, or adapted for welding to the pipe at the other end)
3. Two gaskets which hold the ball and provide a seal
4. A retaining nut or flange to hold the ball and gaskets in the casing

The end of one of the two pipes being coupled is connected to the joint casing; the end of the other, to the ball. In operation, a ball joint accommodates movements of the pipes by providing a flexible articulation (30° to 40° total angular flex, plus 360° rotation or swivel motion).

## Air Vents

Air pockets in piping prevent or interfere with water circulation. Therefore, all piping must be pitched to vent air into the expansion tank through connections on the radiation equipment, or at high points in the system. The mains are pitched up and away from the boiler or heater (in the direction of flow) at a slope of 1/4 to 1 inch per 10 feet. Either manual or automatic vents must be installed on the radiation equipment, or at high points in the system. If manual vents are used, the air must be vented periodically. Install the minimum number of automatic vents, valves, and fittings, as they are points of air entry or water leakage. (See Figure 15–12.)

## Mechanical Circulators

Centrifugal pumps, usually directly connected, are used for circulators. They are mounted in the return system near the boiler inlet, since the water

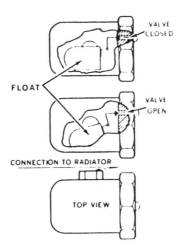

FLOAT

VALVE CLOSED

VALVE OPEN

CONNECTION TO RADIATOR

TOP VIEW

**FIGURE 15-12.**
One type of automatic air vent

temperature is lowest there. The pump is selected on the basis of its ability to circulate the required volume of water against the resistance head of the system as designed. Flow velocities above 4 feet per second are likely to cause disturbing noises in the system, so this figure is frequently considered a maximum value for system design. The required pump capacity is determined from the heating capacity required and the design temperature drop between the supply and the return piping, which is usually 20°F to 30°F. For example: If a capacity of 10,000 Btu is required and a 20°F drop is selected, the pump capacity needed would be 5,000 lbs per hour or about 10 gpm. In most large systems, the pumps run continuously. In small systems operation may be either continuous or intermittent, depending on the type of control equipment used. Many systems use a valved bypass line around the circulator in case of pump failure. Some large systems use a small steam turbine to drive the pump and use the turbine exhaust steam to heat water, especially during mild weather. Steam loss to the atmosphere results under these circumstances.

Pipe size has an important effect on pump power requirements. Friction heads increase almost as the square of flow velocities. A large pipe means lower velocities and reduced power needed to run the pump, but the larger pipe costs more initially. The usual centrifugal circulator has an efficiency of about 60%. The cost of power for a large installation must be determined in advance, and the annual cost for power must be compared with cost of piping of various sizes to insure the most economical combination.

Circulators are usually centrifugal water pumps varying in size from small 1/6 hp booster pumps with a capacity of 5 gpm at 6 or 7 feet head, to large units which pump hundreds of gallons a minute against high heads. They circulate the hot water through the system and inject it into the boiler against boiler pressure.

Many hot water heating pumps use mechanical seals or rely on deep stuffing boxes. Usually, booster pumps are installed by setting the unit in the pipeline as if it were a flanged fitting. They can be installed in either the supply or the return header adjacent to the boiler, but the supply header is more desirable.

Large installations normally accomplish zoning by using individual pumps in each zone return circuit. When a single pump system is used, each return line should be connected to a common, pump return-header through a square head lock. This permits equalization of the return temperatures from the various circuits and thus balances the system. Immersion thermometers in the return and near the pump often help assure system balance.

The piping arrangement and design determines whether pumps operate in a parallel or series manner. In parallel operation, the total flow at the common discharge head equals the sum of the flow capacity of each pump. In series operation, the flow for each pump is identical; the common head is established by adding the pump heads at the identical flow.

# FLOW ADJUSTMENT AND BALANCE

A properly designed hot water heating system is, to a large extent, self-adjusting. However, most have some way to regulate the water flow and thereby adjust the heat delivered by each heating unit and branch circuit during unforeseen conditions. Flow adjustment and balancing is especially important in two-pipe, direct-return systems because of their inherent hydraulic unbalance. The following conditions determine the adjusting method to use.

## Pipe Size Selection

If the pipe is properly sized, a flow distribution can be established which assures that the pressure drop through each circuit is the same when the system operates at the design flow rate. However, since this flow control depends on the pressure drop caused by pipe friction at a certain flow, it may not be adequate for other flow conditions.

## Use of Orifices

Orifices can produce friction drops artificially and therefore balance all circuits for design flow conditions. Generally, this method uses the same design principles as pipe sizing except that it creates the required friction drop by introducing an orifice instead of by reducing the pipe size.

## Use of Throttle Valves

Throttle valves provide a flexible arrangement for adjusting circuit water flow and balancing the circuit.

## Air Venting

Air pockets in piping prevent or interfere with water circulation. Therefore, all piping must be pitched to vent air into the expansion tank through connections on the radiation devices or at high points in the piping system. Mains are pitched up and away from the boiler or radiation device (in the vertical direction) at a slope of 1/4 to 1 inch per 10 feet. Either manual or automatic vents must be installed on radiation devices, or at high points in the piping system. If manual vents are used, the air must be vented periodically. Install the minimum number of automatic vents, valves, and fittings, as they are points of air entrance or water leakage.

## System Balancing

To determine whether hot water systems require balancing, measure the space temperatures with room thermometers, or determine the water temperature drops through the radiation devices with thermometers installed in the piping, or with surface contact thermometers. If the temperature drop method is used, the capacity of the radiation devices should be adequately matched to the heat load demand.

**Procedure for System Balancing.**    To adjust the heat distribution of a hot water heating system, proceed as follows:

1. Prepare a worksheet to record all pertinent information, such as, building description, zone, date, equipment data, periodic readings,

space temperature readings, and supply and return temperature of the heating zone or heating units.

2. To eliminate outside influence, balance the system on an overcast day or at night when heat-gain conditions are minimum. However, the outside temperature should be low enough to require at least 50% of the system's heating capacity to maintain the desired inside temperature.

3. Place the system in service.

4. Open all valves, adjusting elements, and dampers for the regulation of air circulation, etc.

5. Make sure that no automatic devices that control the flow of water or the capacity of any unit are operating.

6. Close all doors or connecting openings between rooms.

7. Wait for the system to reach normal equilibrium, then take a complete set of temperature readings throughout the system. (Thermal equilibrium is obtained when successive temperature readings are approximately the same.)

8. Make an initial adjustment of flow control devices according to the registered readings obtained in step 7 above. If adjustment is by space temperature readings, obtain the designed indoor temperature. If the water temperature drop method is used, obtain equal temperature through all units or zones.

9. Take a new set of temperature readings, after a new thermal equilibrium has been established throughout the building.

10. Continue the adjustments of the flow-regulating devices until the desired conditions are obtained.

11. When the system has been satisfactorily adjusted, mark the position of each of the flow-regulating devices (flow cocks, valves, etc.). Then, if the position of the flow-control fittings is disturbed later by accident, the proper flow to the heating units can be reestablished.

# REVIEW QUESTIONS

1. Name the two classifications of low-temperature hot water systems.

2. What causes the water to flow in a gravity system?

3. In what type of water system does the main line size remain the same throughout the circuit?

4. In what type of hot water system does the main line change in size throughout the system?

5. How are two-pipe systems classified?

6. Define a balanced hot water system.

7. In the open-tank hot water system where is the expansion tank located?

8. What is the purpose of the expansion tank in a hot water system?

9. When is a downfeed system used?

10. What will happen if the expansion tank is too small?

11. What is the purpose of the pressure-reducing valve?

12. Why must all hot water heating systems have a pressure relief valve?

13. What device is used to prevent gravitational flow when the circulating pump is off?

14. What is the purpose of the check valve installed in the make-up water line to a hot water boiler?

15. What device is used to regulate the flow rate to control the system temperature?

16. When used in combination with expansion joints, what is the purpose of anchors, hangers, and supports?

17. Name the five types of expansion joints.

18. What shapes do expansion loops usually form in the pipeline?

19. What is the purpose of air vents in a hot water system?

20. Above what water flow velocity are noises likely to occur in a hot water system?

21. How does the flow velocity of water through a system affect the friction load?

22. Why is the flow adjustment and balancing especially important in two-pipe, direct-return systems?

23. How may artificial friction drops be placed in a system?

24. What is the purpose of throttling valves?

25. How should all piping be installed for air venting purposes?

# CHAPTER 16

# STEAM HEATING SYSTEMS

## Learning Objectives

The objectives of this chapter are to introduce the reader to:

- Pressure ranges used in steam boiler systems
- Piping arrangements used in steam boiler applications
- Components used in a steam boiler application
- Types of systems used in steam boiler applications.

A central boiler plant is an assembly of coordinated equipment used to supply the heat needed to meet the job load requirements. Heat is generally delivered as steam and is used to accomplish the desired process.

The boiler is the main unit of a central boiler plant. Every other piece of equipment is subordinate to the boiler. Its function is to boil water at the required temperature to produce steam.

# PRESSURE RANGES

Steam-generating systems are classified according to their piping arrangement, accessories used, method of returning the condensate to the boiler, method of expelling air from the system, or type of control used. The successful operation of a steam-generating system consists of generating steam in a sufficient quantity to equalize the losses at maximum efficiency; expelling entrapped air; and returning all condensate to the boiler rapidly.

Steam cannot enter a space filled with air or water at a pressure which is equal to the supply steam pressure. Therefore, it is important to eliminate the air and to remove the water from the steam distribution system.

The steam boiler operates on the same principle as a closed container of boiling water: Steam formed in the boiling process tends to push against the sides of the vessel. Because of this downward pressure on the surface of the water, a temperature in excess of 212°F is required to make the water boil. The higher temperature is obtained simply by increasing the supply of heat to the water. Bear in mind, therefore, that an increase in pressure over the water means an increase in the boiling point temperature of the water.

Steam has different uses at different pressures, as follows:

1. Low Pressure. Steam at low pressures ranging from 0 to 15 psig is used for space heating, cooking, and distribution within a building.

2. Medium Pressure. Steam pressures within 25 to 50 psig are used for industrial, shop, and warehouse heating, sterilization, and cooking.

3. High Pressure. Steam at 100 psig is used for industrial purposes and for outside distribution systems. Outside distribution pressures can be used, without reduction, on certain types of equipment. (See Figure 16–1.) In an outside distribution system, steam is generated at 100 psig in a central boiler plant and then distributed to the different consuming stations. The diagram in Figure 16–1 shows high-pressure

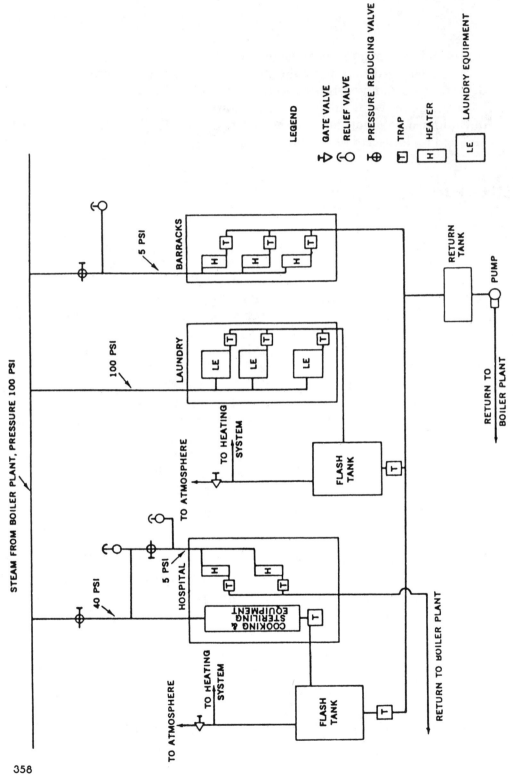

358

**FIGURE 16-1.**
Flow diagram of a steam distribution system

steam being converted to medium-pressure (40 psig) steam for use in cooking and sterilizing equipment; it also shows steam being reduced to low-pressure (5 psig) steam for space-heating systems. High-pressure steam is used directly, without reduction, for some industrial laundry equipment.

## Sources of Steam

Steam can be obtained from several sources. These are discussed below.

**Central Heating Stations.**   For some purposes, high-pressure steam from central heating plants can be used directly without pressure reduction.

**Pressure-Reducing Stations.**   Pressure-reducing stations are used when low- or medium-pressure steam is required, and a source of high-pressure steam is available.

**Converters.**   Low- and medium-pressure steam can be obtained from a converter, when either high-pressure steam, or high-temperature water circulates through coils or tubes submerged in water stored in the converter shell. The heat transmitted through the tube wall converts part of the stored water to low- or medium-pressure steam for space heating or other uses. Feed water is supplied to the shell to replace that converted into steam. A relief valve installed in the shell limits the operating steam pressure to safe values. However, when high-pressure steam is used as the hotter medium, it usually remains static in the shell, while the cooler medium flows through the tubes.

**Low-Pressure Heating Steam Boilers.**   A common source of low-pressure steam is the low-pressure heating steam boiler.

# PIPING ARRANGEMENT

When an exterior steam distribution system is used it may be further divided into underground and aboveground distribution systems.

## Underground Systems

The major types of underground steam distribution systems are the conduit and utilidor types of systems. These systems are generally installed only in permanent installations because of their high cost of installation.

**Conduit Type.**   In the conduit type of steam distribution system, the pipe is installed inside a conduit that is usually buried in the ground below the frostline. The frostline is the lowest depth that the ground freezes during the coldest part of winter. The pipe used for steam is black steel pipe, which is not as strong as that required for the condensate return lines. The conduit and insulation serve to protect and insulate the steam pipe. (See Figure 16–2.) The conduit must be strong enough to withstand the pressure of the earth and the usual additional loads that are imposed upon it.

Several types of materials and various designs are used in the manufacturing of conduit. Common types of conduit are constructed of masonry cement, galvanized iron, and steel. The conduit is generally sealed with asphaltic tar or some other type of sealer to prevent water from getting into the insulation and causing it to deteriorate. Insulation may be attached directly to the pipe, it may be attached to the inner surface of the conduit, or it may be in loose form, packed between the pipe and the conduit.

The bottom of the trench for the conduit should be filled with course gravel or broken rock to provide support and adequate water drainage. When allowed to collect, the water seeps into the conduit through porous openings in the sealer. This wets the insulation and causes it to lose much of its insulating value. Manholes are required at intervals along the line to house the necessary valves, traps, and expansion joints. (See Figure 16–3.)

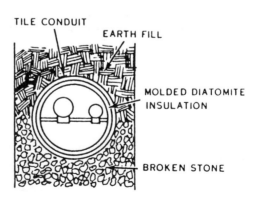

**FIGURE 16-2.**
A steam distribution conduit

TILE CONDUIT

EARTH FILL

MOLDED DIATOMITE
INSULATION

BROKEN STONE

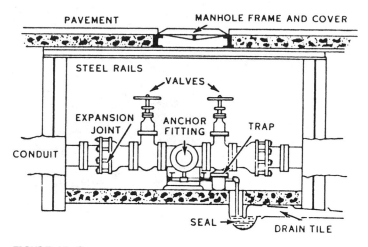

**FIGURE 16-3.**
A typical manhole

**Utilidor Type.** The utilidors or tunnels of the utilidor type of system are constructed of brick or concrete. The size and shape of the utilidor usually depends upon the number of distribution pipes to be accommodated and the depth that the utilidor must go into the ground. Manholes, sometimes doors, are installed to provide access to the utilidor (tunnel). (See Figure 16-4.) The utilidor is usually constructed so that the steam and condensate return lines can be laid along one side of the tunnel on pipe hangers or anchors. This is usually done with the type of pipe hanger that has rollers to allow free movement, which is required because of the expansion and contraction of the pipe. The other side of the utilidor should be a walkway that provides easy access to the lines for inspection and maintenance.

## Aboveground Systems

Aboveground steam distribution systems are further divided into overhead and surface systems.

**Overhead Distribution Systems.** Overhead steam distribution systems are usually used in temporary installations; however, they are sometimes used in permanent installations. The main drawback to this type of distribution system is the high cost of maintaining it. These overhead systems are simpler in many respects to underground distribution systems. They require valves,

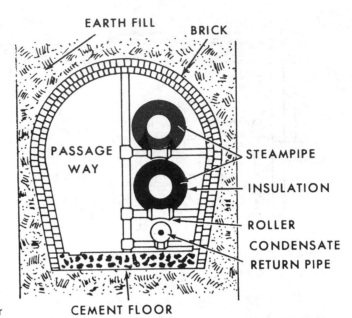

**FIGURE 16-4.**
A typical utilidor

traps, provision for pipe expansion and contraction, and insulated pipes. The main difference is that the steam distribution and condensate return piping are all supported on pipe hangers from poles instead of being buried underground. (See Figure 16–5.)

**Surface Distribution Systems.**   In some cases, steam and condensate lines are laid in a conduit along the surface of the ground. Surface distribution systems, however, are not as common as the overhead and underground

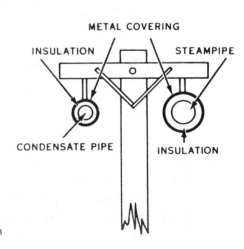

**FIGURE 16-5.**
Overhead steam distribution system

types of distribution systems. Surface distribution systems have about the same requirements as the overhead and the underground systems, namely, traps, valves, pipe hangers to hold the pipes in place, and provisions for pipe expansion and contraction. Sometimes an expansion loop, formed by a loop of pipe, is used instead of an expansion joint to provide for pipe expansion.

## Maintenance

The maintenance required for exterior distribution systems normally consist of inspecting, repairing, and replacing insulation, traps, valves, pipe hangers, expansion joints, conduit, utilidors, and aluminum or galvanized metal coverings used on aboveground distribution systems. The maintenance required on conduit and utilidors consists of keeping the materials of which they are constructed from being damaged, and of insuring that water is kept out of the tunnels and pipes. The maintenance required on outside metal coverings is about the same as that for the conduit and utilidors. Information about the maintenance of traps, expansion joints, and pipe hangers will be given later in this chapter in the discussion of system components.

# EXPANSION JOINTS

Because of the limitations of the supports, or the space or extra length of pipe required, it is sometimes inconvenient to compensate for expansion of a line with special bends and loops. As an alternative, expansion joints may be used. Two general types of expansion joints are in common use; they are ordinarily classed as the slip and the bellows joints. The number of expansion joints installed in a line depends on the amount and direction of expansion and the amount of expansion permitted by each joint. Table 16–1 gives the expansion of pipe per 100 feet for various materials and at various temperatures.

## Slip Expansion Joint

The slip type of expansion joint consists of an outer casing or body which is anchored, a sliding tube which fits into the body, and a means of

**TABLE 16-1**
Expansion of Pipe Per 100 Feet of Length for Temperature Shown

| Temperature, Degrees F | Material | | |
|---|---|---|---|
| | Steel | Wrought Iron | Copper |
| 0 | 0 | 0 | 0 |
| 10 | 0.075 | 0.078 | 0.111 |
| 20 | 0.149 | 0.156 | 0.222 |
| 30 | 0.224 | 0.235 | 0.333 |
| 40 | 0.299 | 0.313 | 0.444 |
| 50 | 0.374 | 0.391 | 0.556 |
| 60 | 0.449 | 0.470 | 0.668 |
| 70 | 0.525 | 0.549 | 0.780 |
| 80 | 0.601 | 0.629 | 0.893 |
| 90 | 0.678 | 0.709 | 1.006 |
| 100 | 0.755 | 0.791 | 1.119 |
| 110 | 0.831 | 0.871 | 1.233 |
| 120 | 0.909 | 0.952 | 1.346 |
| 130 | 0.987 | 1.003 | 1.460 |
| 140 | 1.066 | 1.115 | 1.575 |
| 150 | 1.145 | 1.198 | 1.690 |
| 160 | 1.224 | 1.281 | 1.805 |
| 170 | 1.304 | 1.364 | 1.919 |
| 180 | 1.384 | 1.447 | 2.035 |
| 190 | 1.464 | 1.532 | 2.152 |
| 200 | 1.545 | 1.616 | 2.268 |
| 210 | 1.626 | 1.701 | 2.384 |
| 220 | 1.708 | 1.786 | 2.501 |
| 230 | 1.791 | 1.871 | 2.618 |
| 240 | 1.872 | 1.957 | 2.736 |
| 250 | 1.955 | 2.044 | 2.854 |
| 260 | 2.038 | 2.130 | 2.971 |
| 270 | 2.132 | 2.218 | 2.089 |
| 280 | 2.207 | 2.305 | 3.208 |
| 290 | 2.291 | 2.393 | 3.327 |
| 300 | 2.376 | 2.481 | 3.446 |
| 310 | 2.460 | 2.570 | 3.565 |
| 320 | 2.547 | 2.659 | 3.685 |
| 330 | 2.632 | 2.748 | 3.805 |
| 340 | 2.718 | 2.838 | 3.926 |
| 350 | 2.805 | 2.927 | 4.050 |
| 360 | 2.892 | 3.017 | 4.167 |
| 370 | 2.980 | 3.108 | 4.289 |
| 380 | 3.069 | 3.199 | 4.411 |

| Temperature, Degrees F | Material | | |
|---|---|---|---|
| | Steel | Wrought Iron | Copper |
| 390 | 3.156 | 3.291 | 4.532 |
| 400 | 3.245 | 3.383 | 4.653 |
| 410 | 3.334 | 3.476 | 4.777 |
| 420 | 3.423 | 3.569 | 4.899 |
| 430 | 3.513 | 3.662 | 5.023 |
| 440 | 3.603 | 3.756 | 5.145 |
| 450 | 3.695 | 3.850 | 5.269 |
| 460 | 3.785 | 3.945 | 5.394 |
| 470 | 3.874 | 4.040 | 5.519 |
| 480 | 3.962 | 4.135 | 5.643 |
| 490 | 4.055 | 4.231 | 5.767 |
| 500 | 4.148 | 4.327 | 5.892 |
| 520 | 4.334 | 4.520 | 6.144 |
| 540 | 4.524 | 4.715 | 6.396 |
| 560 | 4.174 | 4.911 | 6.650 |
| 580 | 4.903 | 5.109 | 6.905 |
| 600 | 5.096 | 5.309 | 7.160 |
| 620 | 5.291 | 5.510 | 7.417 |
| 640 | 5.486 | 5.713 | 7.677 |
| 660 | 5.583 | 5.917 | 7.938 |
| 680 | 5.882 | 6.122 | 8.197 |
| 700 | 6.083 | 6.351 | 8.460 |
| 720 | 6.284 | 6.539 | 8.722 |
| 740 | 6.488 | 6.749 | 8.988 |
| 760 | 6.692 | 6.961 | 9.252 |
| 780 | 6.899 | 7.175 | 9.519 |
| 800 | 7.102 | 7.384 | 9.783 |
| 820 | 7.318 | 7.607 | 10.056 |
| 840 | 7.529 | 7.825 | 10.327 |
| 860 | 7.741 | 8.045 | 10.598 |
| 880 | 7.956 | 8.266 | 10.872 |
| 900 | 8.172 | 8.489 | 11.144 |
| 920 | 8.389 | 8.713 | 11.420 |
| 940 | 8.608 | 8.939 | 11.696 |
| 960 | 8.830 | 9.167 | 11.973 |
| 980 | 9.052 | 9.396 | 12.253 |
| 1,000 | 9.275 | 9.627 | 12.532 |
| 1,100 | 10.042 | 10.804 | 13.950 |
| 1,200 | 11.598 | 12.020 | 15.397 |

preventing leakage between the inner and the outer sections. (See Figure 16–6.)

Figure 16–7 is an example of a double joint in which a plastic type of packing is used. The plunger is backed out; packing is forced in with a pressure gun similar to that used for greasing automobiles; and the plungers are then screwed down. This joint may be applied with welded or flanged construction. Several types of similar joints use ordinary packing adjusted by means of a gland. They may be either the outside guided type or the inside guided type of joints. The pipeline must be held in alignment if this type of joint is to function properly. Expansion or contraction moves in the inner sleeve in the main anchored casing or body. Stops must be provided in these joints to prevent them from pulling apart.

## Bellows Expansion Joint

The expansion in the bellows type of expansion joint is taken care of by the flexing of a metal bellows. When installed in the line, the joint consists of a corrugated, thin-walled copper tube which is clamped between the flanges. (See Figure 16–8.) The rings help to keep the corrugations in the joint under high pressures. This joint does not usually have a safety stop.

Another type of joint uses a stainless steel multidisc bellows. (See Fig-

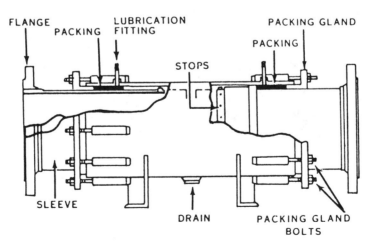

**FIGURE 16-6.**
Slip type of expansion joint

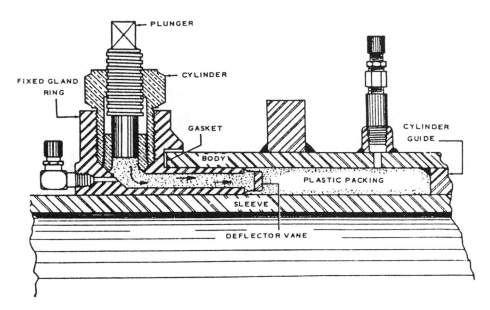

**FIGURE 16-7.**
Slip type of expansion joint with plastic packing

ure 16–9.) This joint has an internal sleeve which acts as a stop to limit the flexing of the discs and to limit the maximum flow of steam through the bellows if the discs should rupture. Both of these joints can compensate for a small amount of misalignment, but the pipe should be supported and guided in such a way that misalignment is reduced to a minimum. It is best that misalignment be prevented altogether if possible.

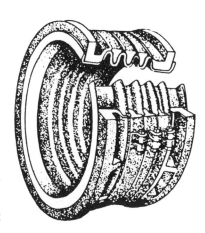

**FIGURE 16-8.**
Corrugated bellows type of expansion joint

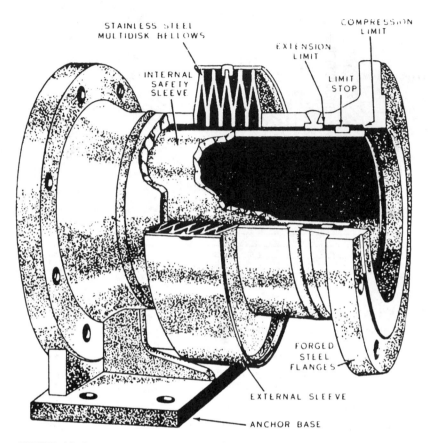

**FIGURE 16-9.**
Multidisc bellows type of expansion joint

## Swing Joint

The swing or swivel joint is most often used to allow for expansion. It is better adapted to screwed joints. When it is used with welded elbows, the swing introduces torsional strains in the elbows and the swing piece.

## Expansion Loops

The expansion loop absorbs the expansion by introducing U or Z loops in the pipe.

## Ball Joints

Ball joints are often used instead of expansion loops because they require less space and material. Ball joints have four basic parts:

1. The body casing to hold the gaskets and the ball
2. The ball (a hollow fitting shaped like a ball at one end, which is inside the casing, and threaded, flanged, or adapted for welding to the pipe at the other end)
3. Two gaskets which hold the ball and provide the seal
4. A retaining nut or flange to hold the ball and gaskets in the casing

The end of one of the two pipes being coupled is connected to the joint casing, and the end of the other is connected to the ball. In operation, a ball joint accommodates movements of the pipe by providing a flexible articulation (30° to 40° total angular flex, plus 360° rotation or swivel motion).

## Inspection and Maintenance of Expansion Joints

A description of the inspection and maintenance procedures for expansion joints is given below.

**Slip Type.**   Slip type expansion joints must be kept properly aligned, adequately packed, within the proper limit of travel, and thoroughly cleaned and lubricated. Adjust or replace packing, as required, to prevent leaks and to assure a free working joint. Lubricate every six months, using the proper grease for the type of joint and service conditions. Once a year, check the flange-to-flange distance of slip joints, first when cold and then when hot, to make sure that travel is within the limits shown in the manufacturer's data. A change in slip travel usually indicates a shift in anchorage or pipe guide. Locate and correct the difficulty. Also inspect annually for signs of erosion, corrosion, wear, deposits, and binding. Repair or replace defective parts as required.

**Bellows Type.**   Annually, check bellows joints for misalignment, fatigue, corrosion, and erosion; note the amount of travel between cold and hot conditions. If the joint fails, replace the bellows section.

**Expansion Loops.**    Expansion loops require no specific maintenance except inspection for alignment.

**Ball Joints.**    See that the joint is adequately packed. Adjust or replace gaskets, as required, to prevent leaks and obtain a free working joint. Refer to the manufacturer's instructions.

**Swing Joints.**    These joints require only the normal maintenance for pipe and fittings.

# TRAPS AND STRAINERS

A steam strap is a device used to drain water from a steam pipe, separator, radiator, kettle, sterilizer, or other radiation device without allowing the steam to escape. The trap is an important piece of equipment; without it, the water is not removed, and the associated apparatus will not heat. If steam is permitted to blow through the trap, heat is lost. The common difficulties experienced with traps are: cut seats caused by throttling the flow; air binding; stoppage by foreign material; and worn pins or bearings. Seats of the float type of trap are more liable to be cut than those of the intermittent discharge type. This is particularly true for float traps operating at low flows. Seats which are badly cut or worn permit the loss of steam. An airbound trap is inoperative. Worn bearings and pins may cause the trap to stick and make it inoperative. Traps may be classified according to their operation as: thermostatic, float, bucket, impulse, and throttling traps.

## Thermostatic Trap

Thermostatic traps are widely used on radiators. (See Figure 16–10.) The sylphon bellows contains a fluid which expands and vaporizes when heated. This action builds up a pressure inside the bellows, causing it to elongate and close the valve. When water collects around the sylphon and cools it slightly, the sylphon contracts, opening the valve and letting the water escape. As water is forced out, steam comes into contact with the bellows, causing it to close the valve, thus preventing the escape of steam. The sylphon and the lower valve seat can be removed from the trap without disturbing any of the pipe work.

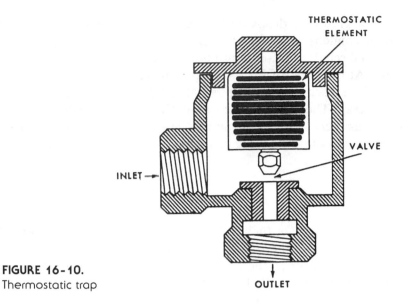

**FIGURE 16-10.**
Thermostatic trap

## Float and Thermostatic Trap

In the combination float and thermostatic trap, the thermostatic trap normally vents the air, but it can also discharge water if the capacity of the float valve is exceeded. Water enters the trap and raises the float, carrying the vertical valve with it. (See Figure 16–11.) This action opens the valve and permits the water to discharge; as the water level in the chamber drops,

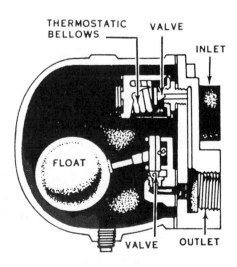

**FIGURE 16-11.**
Float and thermostatic trap

the float gradually drops and closes the valve. Normally, with a constant flow of condensate to the trap, the level of water in the trap is stable and the trap discharges water continuously at the same rate that the condensate enters. At low rates of flow, the level of water in the trap is low, and throttling of the discharge causes the seat to be cut badly—a disadvantage of this type of trap. The thermostatic trap in the top of the chamber remains closed as long as there is steam around it. This trap can be opened for inspection or repair without disturbing any pipe connections.

## Bucket Traps

Bucket traps get their name from the fact that the operating element is a small bucket. The trap may be made with the open end of the bucket either at the top or at the bottom. When the open end of the bucket is at the top, the device is called a bucket trap; when the open end is at the bottom it is called an inverted bucket trap.

**Inverted Bucket Traps.**   In operation, the steam under the bucket slowly condenses and permits water to fill the bucket until it sinks, opening the valve and discharging the water. (See Figures 16–12 and 16–13.) After all the water is removed from the inlet line, steam discharges into the trap under the bucket, forcing the water out and causing the bucket to rise and close the valve. As can be seen, the trap discharges only to the level of the valve, thus leaving the trap immersed at all times. To prevent the trap from

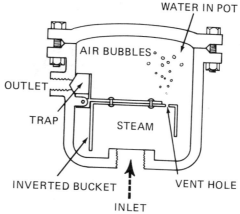

**FIGURE 16-12.**
Inverted bucket trap with trap closed

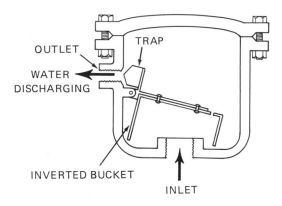

**FIGURE 16-13.**
Inverted bucket trap discharging water

becoming airbound, it is provided with an air vent hole. This trap can be disassembled for inspection and repair without disturbing the piping.

The construction and operation of bucket traps is shown in Figures 16–14, 16–15, 16–16. The steam and condensate enter the body of the trap (Figure 16–14) in which an empty bucket floats. The water flows over the edge

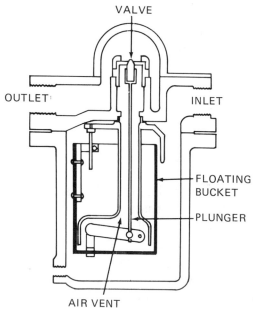

**FIGURE 16-14.**
Bucket trap with trap closed

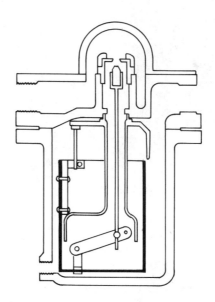

**FIGURE 16-15.**
Bucket trap discharging water

of the bucket and starts to fill it. When enough water has entered, the bucket sinks, allowing the water to be forced out. (See Figure 16–15.) When enough water has been forced out, the bucket rises, closing the valve and leaving the column from the bucket bottom to the valve full of water. If air is present, it is drawn into the column as shown in Figure 16–16, and the water in the column drops into the bucket as in Figure 16–15. The air is discharged the next time that the valve opens.

## Impulse Trap

In the impulse trap, the flashing action produced by a pressure drop in the hot condensate governs the movement of a valve by changing the pressure in a control chamber above the valve. (See Figure 16–17.) In operation, condensate builds up pressure below the control disc, lifting the valve like a piston. Air and condensate discharge, and a small portion of the flow (control flow) moves up around the disc to the lower pressure-control chamber. The pressure in this chamber remains low while the control flow discharges through the orifice in the valve body to the trap outlet side. When condensate reaches near-steam temperature, the reduced pressure in

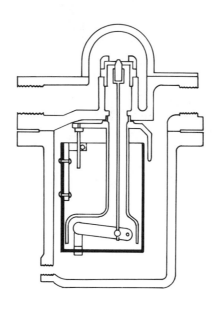

**FIGURE 16-16.**
Bucket trap after discharging water

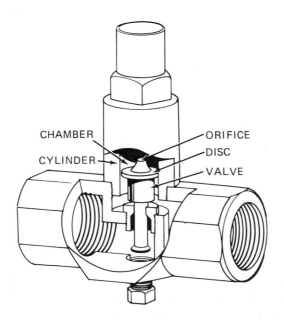

CHAMBER

CYLINDER

ORIFICE

DISC

VALVE

**FIGURE 16-17.**
Impulse trap

the control chamber causes part of the control flow to flash into steam. The increased volume in the control chamber chokes off some of the flow through the control orifice to build up pressure in the chamber. The valve closes and shuts off all discharge except the small amount flowing through the control orifice. Impulse traps can be used to drain condensate from steam mains, unit heaters, laundry equipment, sterilizers, and other equipment in which the pressure at the trap outlet is 25% less than the inlet pressure.

## Throttling Trap

A throttling trap operates on the principle that the flow of water through an orifice decreases as its temperature approaches that of the steam used. This trap has no moving parts and the rate of flow through it can be adjusted by raising or lowering the stem, which fits a tapered V-seat. (See Figure 16–18.) When the stem is properly adjusted, the condensate, which is slightly cooler than the steam, enters the chamber from the line, travels up through the baffle passage, and is discharged through the orifices. If this discharge is at a rate higher than that at which it enters the trap, the level in the inlet chamber falls until it permits a little steam to enter the baffle passage. The steam going through this passage heats the condensate to a temperature approaching that of the steam. As the temperature increases, the amount of water flashing into steam, and hence the volume of the steam water mixture handled by the orifice, permits the level of condensate in the inlet chamber to rise until the hotter water in the baffle passage has been completely discharged and replaced with slightly cooled water. The cycle is then repeated. The orifice vents air from the trap; otherwise, steam would be excluded and the condensate would cool. This condition results in a high flow through the orifice, which drops the water level and permits the air to enter the baffle passage and escape.

## Strainers

Pipeline strainers are used in both water and steam lines to prevent foreign matter from entering and clogging the equipment. (See Figure 16–19.) A strainer is used ahead of a trap or a pressure-reducing valve to prevent fouling of the orifice or operating mechanism. Ordinarily, the screen of the strainer is thin sheet metal, usually bronze, monel, or stainless steel. The strainer can be removed for cleaning or it may be arranged with a valve to permit blowing out the foreign material. The strainer must be blown out

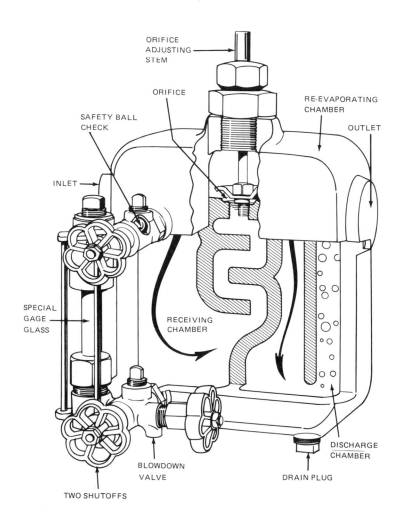

**FIGURE 16-18.**
Throttling trap

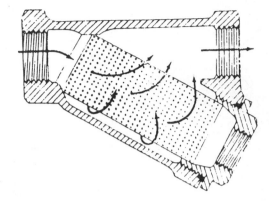

**FIGURE 16-19.**
Pipeline strainer

regularly or examined and cleaned to prevent the flow from being reduced or stopped. If the flow is obstructed, the screen may break and permit foreign material to pass through.

# CONDENSATE RETURN

Steam distribution systems are also classified according to the method used for returning the condensate to the boiler. The classification depends upon whether the condensate is returned by gravity or by mechanical means.

In the gravity system the condensate is returned because of a static head of water in the return pipe. In this system all radiation must be above the boiler waterline. In the mechanical system the condensate flows by gravity to a receiver and is then forced into the boiler against pressure. The main difference between mechanical and gravity systems is that radiation in the mechanical system can be located below the boiler waterline, as long as there is a low spot to which the condensate can be drained and from which it can be pumped into the boiler, feed water heater, or surge tank.

## Mechanical Return of Condensate

In gravity return systems, all units must be located high enough above the boiler waterline to produce a gravity flow of condensate toward the boiler. Many times this condition cannot be met, and the condensate must be returned by mechanical means. Two methods for mechanically returning condensate are used: (1) alternating return trap systems, and (2) pumped return systems.

**Alternating Return Trap System.**    The two-pipe, vapor, alternating return trap system, as its name implies, alternately fills and dumps. (See Figure 16–20.) It returns condensate to the boiler by a mechanical, alternating return trap instead of gravity. The alternating return trap consists of a vessel with a float which, by linkages, controls two valves simultaneously so that one is closed when the other is open. One valve opens to the atmosphere; the other is connected to the steam header. The bottom of the vessel is connected to the wet return. In operation, when the float is down, the valve connected to the steam header is closed and the other is open. As the condensate returns, it goes through the first check valve and rises into the return trap, which is normally located 18 inches above the boiler waterline. The float starts to rise and, when the water reaches a certain level in the trap, the air vent closes and the steam valve opens. This equalizes the trap and boiler pressures, permitting the water to flow by gravity from the trap, through the boiler check valve, and into the boiler. The float then returns the trap to the normal venting position, ready for the next flow of return water.

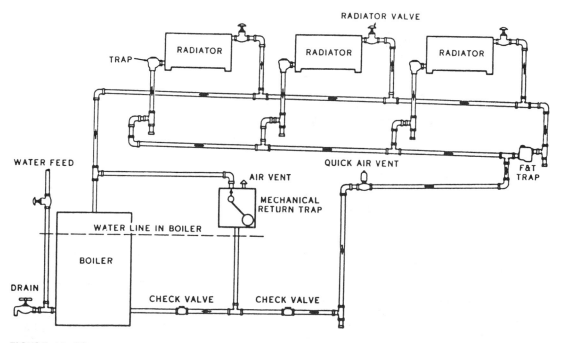

**FIGURE 16-20.**
Two-pipe, vapor, alternating return trap systems

## Condensate Pump Return System

In this type of system, a centrifugal pump forces the condensate into the boiler under atmospheric or higher pressure.

**Two-Pipe Condensate Pump, Return System.**    The typical two-pipe condensate pump, return system, is a low-pressure system that returns condensate to the boiler with a condensate pump. (See Figure 16–21.) Thermostatic traps in the radiator return connections remove air and condensate from the equipment. Radiators, convectors, and small heating equipment can usually be drained with thermostatic traps. However, float-and-thermostatic traps are needed to drain unit, blast, and coil heaters and other equipment that accumulates large volumes of condensate. A check valve at the pump discharge between the pump and the boiler prevents a reverse flow; a gate valve provides for isolation. The main return discharges by gravity into the receiver tank; a valved connection permits system drainage; and a vent in the receiver vents air and flash steam.

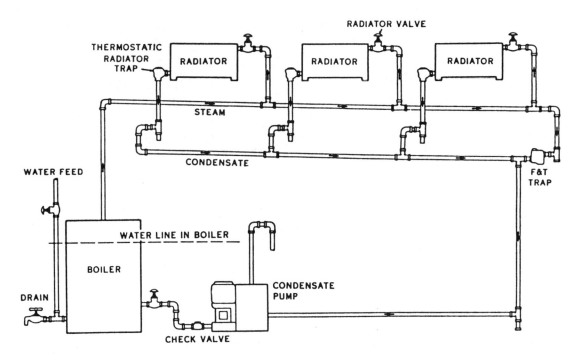

**FIGURE 16-21.**
Two-pipe condensate pump return system

A typical condensate pumping unit consists of a pump, motor, receiver, and a float control. (See Figure 16–22.) The pump operates intermittently under the action of the float-controlled switch.

## Vacuum Pump Return System

In this system a vaccum pump creates a vacuum in the return piping. (See Figure 16–23.) The return outlet of each heating unit in this system is fitted with a thermostatic trap which allows both air and condensate to pass, but closes against flow. This system may operate with low pressures in the supply main, but it can also maintain a vacuum in the return piping for all

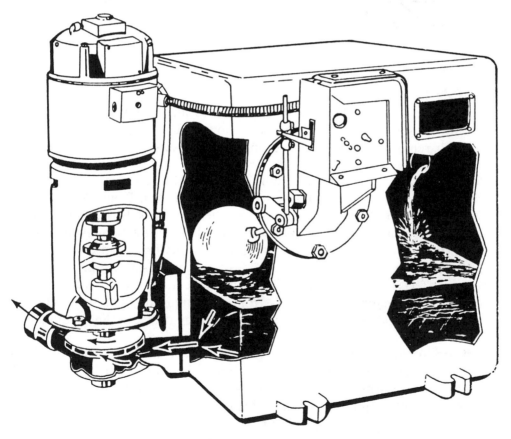

**FIGURE 16-22.**
Condensate pumping unit

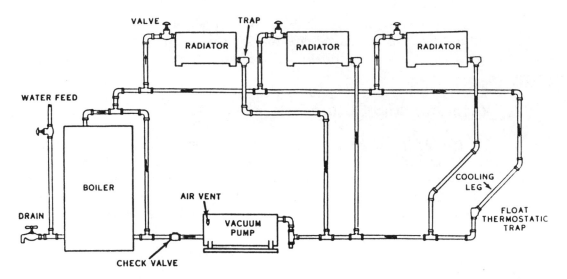

**FIGURE 16-23.**
Two-pipe vacuum system

operating conditions. If a vacuum of 3 to 10 inches is kept in the return system, heating units can fill rapidly at low steam pressures (0 to 2 psig), since air removal does not depend on steam pressure.

**Vacuum Pump.**    The vacuum pump performs the following functions: it withdraws air and water from the system; it separates air from water; it vents air to the atmosphere; and it pumps condensate back to the boiler, feed water heater, or hot well. Usually the pump has both a float switch and a vacuum switch. In this type of system, all connections from the supply to the return side must be made through a trap. Condensate temperature conversely affects the vacuum maintained; the higher the temperature, the lower the possible vacuum. Steam leaks from the supply to the return side will increase the condensate temperature and thereby decrease the vacuum.

**Induced Vacuum.**    When fuel combustion stops, condensation of the steam which fills the heating units can cause an induced vacuum which may prevent drainage of condensate from the heating units and cause a serious deficiency in boiler water supply. An equalizer line is usually installed between the supply main and the return piping to relieve this condition. The equalizer line shown in Figure 16–23 includes a thermostatic trap which remains closed when steam is flowing, but when the steam flow stops, opens to equalize the pressures in the supply and return mains. This permits the condensate to flow back to the pump receiver.

# FEED WATER SYSTEMS AND COMPONENTS

All boilers use a feed water system to provide the water that must be added to the boiler as steam is generated. This water may be condensate that has been returned or make-up water that is needed to replace water lost through leakage during operation of equipment. Feed water systems have been developed to add this water to the boiler when needed. These systems vary from a simple water main supply used with small low-pressure boilers, to the complex system used in high-pressure mutliboiler installations.

The following important advantages accrue when boiler feed water is heated and deaerated:

1. The use of exhaust steam (from pumps, engines, etc.) for heating increases boiler plant efficiency about one percent for every 10 or 12°F increase in boiler feed water temperature. An additional gain occurs when the recovery of condensing steam reduces boiler blowdown requirements. No increase in efficiency results from the use of live steam for feed water heating.

2. Removal of oxygen, carbon dioxide, and other corrosive gases from the boiler feed water (deaeration) prevents corrosion of the ferrous parts of the boiler plant equipment and piping which are normally in contact with the feed water. As a rule, corrosive gases are noncondensable.

3. Introducing hot water into the boiler reduces the thermal shock and strain produced when relatively cold water strikes hot metal parts. The smaller the temperature difference between the feed water and the hot metal parts, the less the strain produced.

4. Reducing the amount of heat required to generate steam increases boiler capacity. When water is fed into the boiler at a temperature of 212°F, the heat required to generate a pound of steam at a pressure of 100 psig is 90% as much as that required for water supplied at 100°F. Therefore, if the same amount of heat is used, more steam will be generated from heated water. There are several types and designs of feed water heaters available, and each unit differs according to the

manufacturer. In general, however, feed water heaters may be divided into three types: open, closed, and economizers.

## Open Feed Water Heater

The open type of feed water heater heats water by direct contact with the steam. (See Figure 16–24.) The heater is usually cylindrical in shape and has an air vent located at the top. The steam enters the side of the heater through an oil separator. The water to be heated, usually condensate return, enters at the top and is distributed over the trays in rainfall fashion. The steam comes in direct contact with the water and is condensed as the water is heated. Oxygen and other gases are liberated to the atmosphere through the vent. Such heaters are provided with a sight glass that indicates the water level (not shown), a drain valve for draining the solid materials which settle to the bottom, and a regulating valve in the supply line. Some heaters have

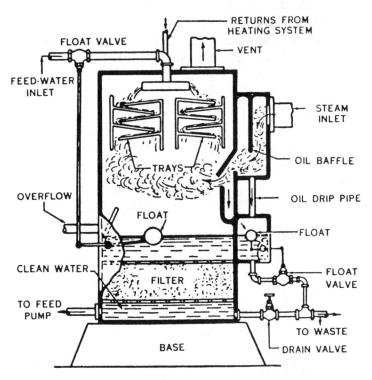

**FIGURE 16-24.**
Open feed water heater

provisions for skimming any foreign matter that collects on the surface of the water in the heater.

## Closed Feed Water Heater

The closed type of feed water heater is constructed so that the steam used to heat the water does not come into direct contact with the water. They are separated from each other by metal surfaces such as tubes. The tank of the heater is supplied with steam, and the water passes through the tubes which are heated by the steam. (See Figure 16–25.) The closed-type feed water heaters are not very common; however, some installations still have them in use.

## Economizer Feed Water Heater

The economizer is another type of feed water heater. Economizers have horizontal tubes arranged in staggered, closely spaced rows. The tubes are placed in the path of the hot flue gases, and the water flows inside the tubes.

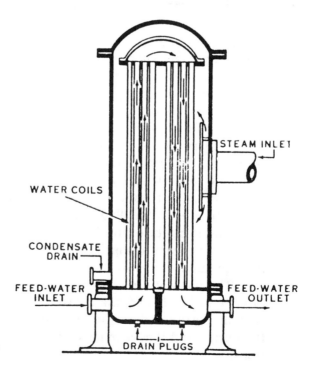

**FIGURE 16-25.**
Closed feed water heater

As the hot flue gases pass around the tube, the water inside is heated. The water can flow either upward or downward through the economizer. However, the upward flow is preferred when there is a possibility of steam being generated in the tubes. When a downward-flow economizer is used and some steam is generated, there is improper circulation through the unit.

The maintenance of feed water heaters and economizers normally includes removing the solid matter that accumulates in the unit; stopping steam and water leaks; and repairing inoperative traps, floats, valves, pumps, and other such equipment.

It is almost impossible to discuss feed water heaters without mentioning deaerators, since they are basically the same type of unit. The only difference between the two is the amount of oxygen each removes from the water. It is generally expected that an open type of heater should remove all oxygen above 0.20 cc (cubic centimeter) per liter; deaerating heaters should remove oxygen in excess of 0.13 ml (milliliter) per liter; and deaerators should remove oxygen in excess of 0.005 ml per liter. The closed-type heaters cannot be classified as deaerators because there is no way to liberate the oxygen.

# TEMPERATURE REGULATION

The purpose of a temperature regulator is to regulate the amount of steam required to maintain the hot water at the desired temperature. The unit consists of a temperature bulb, copper line, diaphragm, spring, temperature adjustment and a steam valve. (See Figure 16–26.)

The bulb and the copper tube are called the capsule and capillary line. They contain a gas which expands or contracts with a change in temperature. The capillary tube is connected to the top of the temperature regulator, which contains a diaphragm (bellows). The diaphragm (bellows) is connected to the valve stem. A spring holds the valve open at low temperatures. When the temperature rises in the water tank, the gas in the temperature bulb expands and forces the diaphragm down, closing the steam valve. The water temperature can be controlled by adjusting the tension of the spring. A steam trap in the steam-heating system returns the condensed steam to the condensate tank.

The hot water tank consists of a temperature gauge that has a range of 40°F to 240°F, and a safety valve or pressure relief valve. The relief valve is set at a pressure that is 10 lbs higher than the operating pressure, and both the setting and the valve must comply with current ASME Code specifications.

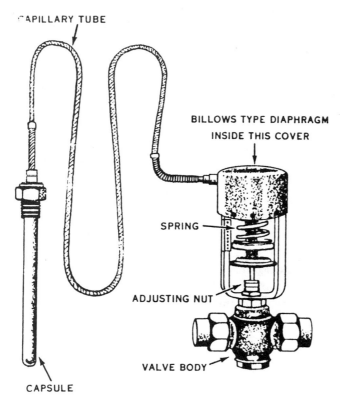

CAPILLARY TUBE

BILLOWS TYPE DIAPHRAGM
INSIDE THIS COVER

SPRING

ADJUSTING NUT

VALVE BODY

CAPSULE

**FIGURE 16-26.**
A temperature-regulating valve

# FEED WATER TANK

In normal plant operation, fluid flows cannot be kept consistently balanced. There are times when the demand for steam from a boiler exceeds the rate at which the water is returned from the system. The reverse can also be true. Feed water tanks compensate for such uneven flows and for the differences between the demand and supply of water at certain times. Water is stored in the tank when the supply exceeds the demand; it is withdrawn to supplement the water feed when the demand is higher than the supply. Feed water tanks installed above the feed water heater supply water to the heater by gravity flow; those that are installed below the feed water heater use a pump to deliver the water to the tank. Tanks usually have water gauges, level controllers, overflows, drains, vents, etc., according to the type of installation.

# FLASH TANKS

Water stores a definite amount of heat energy for each saturation pressure and temperature. When the pressure is suddenly reduced, the heat energy not needed under the new condition is used to evaporate a part of the water. Because this reduction is instantaneous and violent, it is sometimes called "explosive boiling." Explosive boiling occurs when the water at boiler-operation temperature and pressure is suddenly reduced to a lower pressure.

The amount of water flashed into steam by explosive boiling depends on the initial pressure and temperature and the final pressure of the water. For example: When water at a saturation pressure of 100 psig and 338°F temperature is discharged to the atmosphere (0 psig), 13.2 percent of the water is suddenly flashed into steam. If instead the water is discharged to a tank maintaining a pressure of 10 psig, the amount of flash steam formed is reduced to 10.6 percent of the water. As the discharge pressure increases, the amount of water flashed into steam decreases.

## Continuous Blowdown Systems

Flash tanks can be installed as components of boiler blowdown systems. In this type of installation, boiler water with a relatively high temperature and pressure is discharged to the flash tank. (See Figure 16–27.) Part of the water is flashed into steam and can be used for heating or process purposes. The rest of the water is discharged to the sewer, either directly, or after it has passed through heat exchangers to heat the boiler feed water.

## Water Hammer Prevention

Water hammer and other disturbances in the piping system are prevented by using flash tanks to separate the steam from the water. These tanks are usually small and are located close to the traps where the pressure release occurs. (See Figure 16–28.) The steam formed is sometimes used in a lower pressure steam system. Normally, the flash tank has vents to the atmosphere to prevent the trap from discharging against a back pressure.

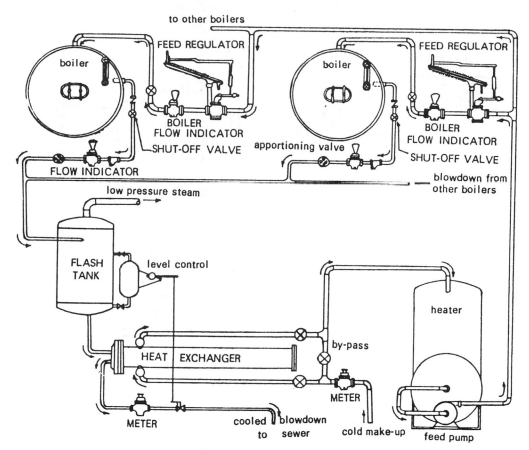

**FIGURE 16-27.**
Continous blowdown system

## Storage Tanks

Storage tanks can be used in central boiler plants for a number of different purposes, for example:

1. To store fuel oil
2. To store lube oil where the settling process will help purify the oil and separate the water
3. To store condensate and receive make-up water from the water treating plant

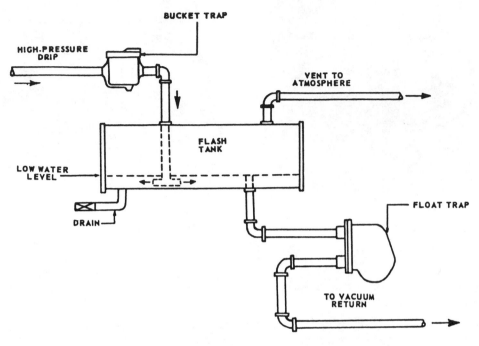

**FIGURE 16-28.**
Flash-tank installation

In each case, the tank accessories vary with the type of service. They may consist of some or all of the following: level gauges, flow controllers, level controllers, overflows, drain valves, vents, pressure gauges, thermometers, measuring rods, relief valves.

# FEED WATER REGULATORS

The purpose of the feed water regulator is to maintain a constant water level, regardless of the boiler load fluctuations. There are three types of regulators in general use:

1. Thermostatic
2. Thermohydraulic
3. Positive displacement

## Thermostatic Water-Level Control

In some big steam feed water systems you will find two large water-level controls, which are commonly referred to as feed water regulators. One of these controls is the thermostatic type. (See Figure 16–29.) It has an expansion generator called a thermostat, which consists of a metal tube mounted on a rigid frame. The tube is connected to the boiler at the steam space and water space.

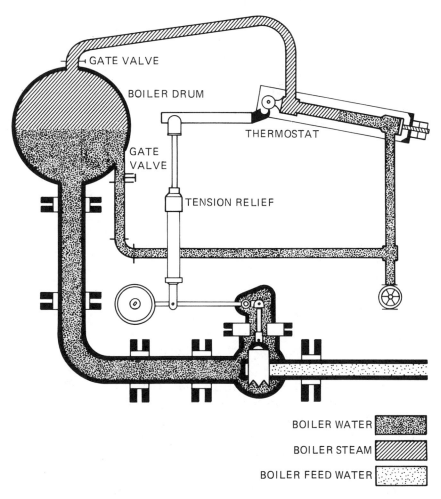

**FIGURE 16-29.**
Thermostatic water-level control

The tube is fastened at one end and is free to move at the other end. Linkage connects the free end of the tube to a valve bellcrank, which in turn, opens and closes the regulating valves. The bellcrank has its fulcrum point attached to the valve.

When the water level in the boiler drops, more steam is allowed to enter the tube and heat it. More steam causes the thermostat to expand and operate the linkage that opens the valve. This allows more feed water to be forced into the boiler. As the water level in the boiler rises, less steam is admitted to the tube of the thermostat; consequently, the tube cools and contracts. The operation of the bellcrank this time closes the regulating valve.

## Thermohydraulic Regulator

The operation of this device is based on the fact that steam occupies a greater volume than the water from which it was formed. The regulator consists of a generator, a diaphragm-operated valve, and connecting pipe and tubing. (See Figure 16–30.) The generator is composed of an inner tube surrounded by an outer tube. The inner tube is connected to the boiler drum or water column and is inclined to keep the normal water level a little above the center of the tube. The outer tube is connected to the diaphragm housing through metal tubing to form a closed system.

Before the generator is started, the regulator valve should be closed, and the generator inner tube should be drained and shut off from the water and steam from the boiler. The generator's closed system (outer tube, tubing, and diaphragm housing) is filled with hot water. When the generator is operating, with the inner tube open to the boiler drum, the water level in the tube corresponds to the boiler water level. Heat from the steam in the upper part of the inner tube raises the temperature of the water surrounding this part of the tube, and converts some of it to steam. The steam pushes water from the outer tube, through the tubing, and into the diaphragm housing. Then pressure, acting on the diaphragm, opens the valve against the force of the spring, and feeds water into the boiler. Water in the generator inner tube raises with the boiler water level and condenses some of the steam. This reduces the heat transfer to the outer tube and lowers the pressure inside the tube. As the pressure is reduced, the spring forces up the diaphragm, water is pushed into the generator, and the valve closes. Fins installed in the generator outer tube radiate some heat, preventing excessive pressures in the closed circuit. (See Figure 16–31.)

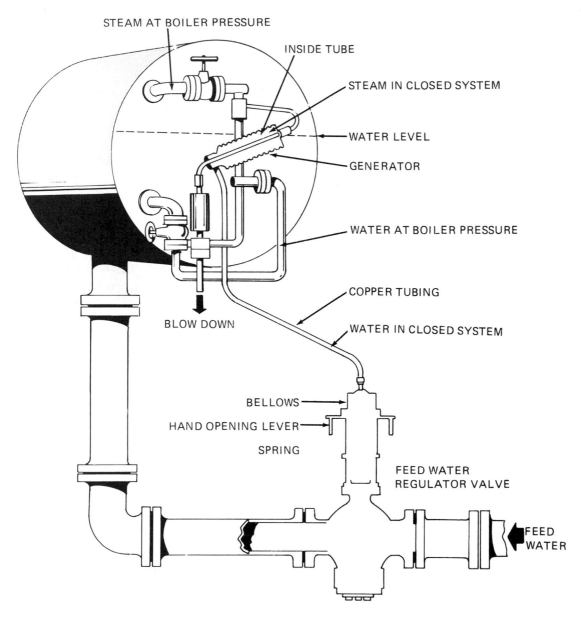

STEAM AT BOILER PRESSURE

INSIDE TUBE

STEAM IN CLOSED SYSTEM

WATER LEVEL

GENERATOR

WATER AT BOILER PRESSURE

COPPER TUBING

BLOW DOWN

WATER IN CLOSED SYSTEM

BELLOWS

HAND OPENING LEVER

SPRING

FEED WATER REGULATOR VALVE

FEED WATER

**FIGURE 16-30.**
Thermohydraulic water-level control

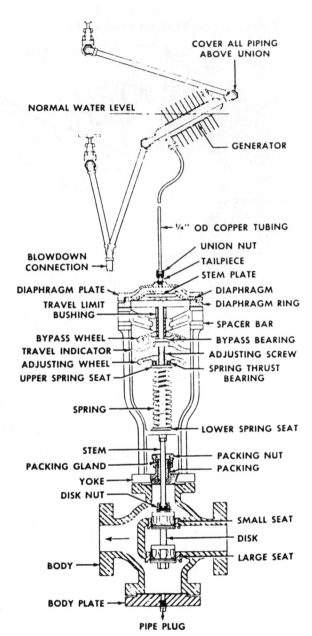

COVER ALL PIPING
ABOVE UNION

NORMAL WATER LEVEL

GENERATOR

¼" OD COPPER TUBING

UNION NUT
TAILPIECE
STEM PLATE
DIAPHRAGM
DIAPHRAGM RING
SPACER BAR
BYPASS BEARING
ADJUSTING SCREW
SPRING THRUST
BEARING

LOWER SPRING SEAT

PACKING NUT
PACKING

SMALL SEAT
DISK
LARGE SEAT

BLOWDOWN
CONNECTION

DIAPHRAGM PLATE
TRAVEL LIMIT
BUSHING

BYPASS WHEEL
TRAVEL INDICATOR
ADJUSTING WHEEL
UPPER SPRING SEAT

SPRING

STEM
PACKING GLAND
YOKE
DISK NUT

BODY

BODY PLATE

PIPE PLUG

**FIGURE 16-31.**
Thermohydraulic feed water
regulator

## Positive Displacement

The float and level regulator is of the positive displacement type. (See Figure 16–32.) The float chamber is connected to the boiler steam space or water space so that its mean water level corresponds to that of the boiler. The feed valve is the balanced type, and there are no stuffing boxes to leak or cause binding.

In operation, the float follows the water level, opening the valve through a suitable system of levers to increase the water flow. The valve and linkage provide gradual changes in the flow of water to maintain an almost constant water level. A small amount of alcohol is introduced into the float and vaporizes with the heat. It builds up sufficient pressure to counteract the boiler pressure on the outside of the float and prevent its collapse. To trace a feed water system, the boiler would be the most likely place to start. Some items to look for while tracing the system are valves leaking, proper installation, and the proper operation of the float.

## Feed Water Control

The boiler feed water control unit is provided to maintain a fixed water level in the boiler. (See Figure 16–33.) It also protects the boiler against low water conditions. This unit consists basically of an enclosed float, and electrical switches. It is connected at the mean water level. (This unit adds water to the boiler by starting the feed water pump.) When the normal water level is reached, the float opens the electrical circuit to the pump. It also provides for ringing a bell when the water in the boiler becomes either dangerously

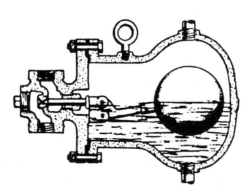

**FIGURE 16-32.**
Cross section of positive displacement type of feed water regulator

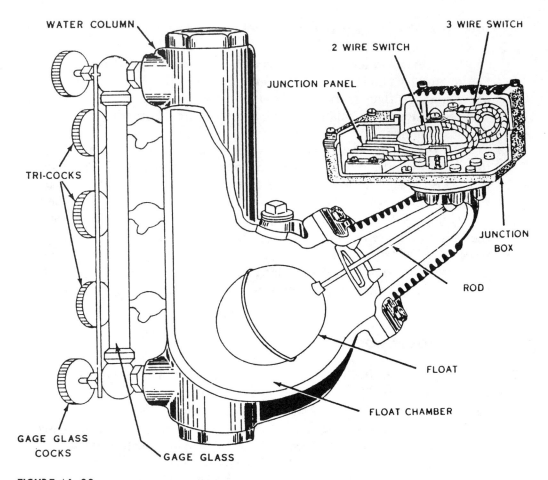

**FIGURE 16-33.**
Cross section of boiler feed water control unit

high or low. The unit also provides for stopping the operation of the fuel-burning equipment when the water level reaches a dangerously low point.

## REVIEW QUESTIONS

1. How is the temperature of the water and steam increased in a boiler?
2. What is the steam pressure range in a low-pressure steam system?
3. What is steam at 100 psig used for?
4. Name the two types of exterior steam distribution systems.

5. What type of systems are the conduit and utilidor distribution systems used in?

6. In what types of installations are overhead distribution systems usually used?

7. What type of expansion joint is best adapted to screwed joints?

8. Why are ball joints often used instead of expansion loops?

9. What type of maintenance is required on expansion loops?

10. What device is used to drain water from a steam system device?

11. What is the purpose of pipeline strainers?

12. What may happen to the strainer screen if it becomes clogged?

13. Name two methods of mechanically returning condensate to the boiler.

14. What type of switches may be used on a vacuum pump?

15. What two sources provide make-up water to a boiler?

16. What is the purpose of the boiler feed water heater?

17. What device is used to control the amount of steam into the feed water heater?

18. Where is the excess water stored during periods of low steam generation?

19. What causes explosive boiling?

20. What device will help to prevent water hammer in a steam system?

21. What device maintains a constant water level in a boiler?

22. Name three types of feed water regulators.

23. When two feed water regulators are used on a boiler, what type will one of them be?

24. On what principle does the thermohydraulic feed water control operate?

25. What is the purpose of the feed water control?

# GLOSSARY

Portions of this section are presented with the permission of the American Boiler Manufacturers Association, Arlington, Virginia.

*Absolute pressure:* A pressure which is above zero gauge. It is equal to the sum of atmospheric and gauge pressures.

*Acid cleaning:* The process of cleaning the interior surfaces of steam-generating units by filling the unit with a diluted acid accompanied by an inhibitor to prevent corrosion, and by subsequently draining, washing, and neutralizing the acid by a further wash of alkaline water.

*Acidity:* This represents the amount of free carbon dioxide, mineral acids, and salts (especially sulfites or iron and aluminum) which hydrolize to give ions in water and is reported as milli-equivalents per liter of acid, or ppm acidity as calcium carbonate, or pH (the measure of hydrogen ion concentration).

*Air vent:* A valved opening in the top of the highest drum of a boiler or pressure vessel for venting air.

*Alkalinity:* This represents the amount of carbonates, bicarbonates, hydroxides and silicates or phosphates in the water and is reported as grains per gallon, or ppm as calcium carbonate.

*Allowable working pressure:* The maximum pressure for which the boiler was designed and constructed; the maximum gauge pressure on a complete boiler; and the basis for the setting of the pressure-relieving devices protecting the boiler.

*Ambient temperature:*   The temperature of the air surrounding the equipment.

*Amine:*   A chemical which is used to control corrosion in condensate systems. Classified as filming or neutralizing.

*Anion exchanger:*   An ion exchange device which removes alkalinity and sulfate when regenerated with salt. It removes sulfate, chloride, and nitrate when regenerated with caustic soda. This process may also remove silica.

*ASME Boiler and Pressure Vessel Code:*   The boiler and pressure vessel code of the American Society of Mechanical Engineers with amendments and interpretations thereto made and approved by the council of the Society.

*Blowdown:*   Removal of a portion of boiler water for the purpose of reducing concentration, or to discharge sludge.

*Blowdown valve:*   A valve which is generally used to continuously regulate the concentration of solids in the boiler water (not a drain valve).

*Blowoff separator:*   A vented and drained container equipped with internal baffles or an apparatus for the purpose of separating moisture from flash steam as it passes through the vessel.

*Blowoff valve:*   A specially designed, manually operated valve which is connected to the boiler for the purpose of reducing the concentration of solids in the boiler water or for draining purposes.

*Boiler:*   A closed vessel in which water is heated, steam is generated, steam is superheated, or any combination thereof, under pressure or vacuum by the application of heat from combustible fuels, electricity, or nuclear energy. The term does not include such facilities of an integral part of a continuous processing unit but shall include fired units of heating or vaporizing liquids other than water where these units are separate from processing systems and are complete within themselves. The different types of boilers are as follows:

*Bent tube:*   A water-tube boiler consisting of two or more drums connected by tubes, practically all of which are bent near the ends to permit attachment to the drum or shell on radial lines.

*Box header:*   A horizontal boiler of the longitudinal or cross drum type consisting of a front and rear inclined rectangular header connected by tubes.

*Cross drum:*   A sectional header or box boiler in which the axis of the horizontal drum is at right angles to the center lines of the tubes in the main bank.

*Firetube:*   A boiler with straight tubes, which are surrounded by water and steam and through which the products of combustion pass.

*Horizontal:* A water-tube boiler in which the main bank of tubes are straight and on a slope of 5 to 15 degrees from the horizontal.

*Horizontal firebox:* A fire-tube boiler with an internal furnace, the rear of which is a tube sheet directly attached to a shell containing tubes. The first-pass bank of tubes is connected between the furnace tube sheet and the rear end. The second-pass bank of tubes, passing over the crown sheet, is connected between the front and rear end closures.

*Horizontal return tubular:* A fire-tube boiler consisting of a shell, with tubes inside the shell attached to both end closures. The products of combustion pass under the bottom half of the shell and return through the tubes.

*Locomotive:* A horizontal fire-tube boiler with an internal furnace, the rear of which is a tube sheet directly attached to a shell containing tubes through which the products of combustion leave the furnace.

*Longitudinal drum:* A sectional boiler or box-header boiler in which the axis on the horizontal drum or drums is parallel to the tubes in a vertical plane.

*Low head:* A bent-tube boiler having three drums with relatively short tubes in a vertical plane.

*Refractory-lined firebox:* A horizontal fire-tube boiler, the front portion of which sets over a refractory or water-cooled refractory furnace. The rear of the boiler shell has an integral or separately connected section containing the first-pass tubes, through which the products of combustion leave the furnace. These products return through the second-pass upper bank of tubes.

*Scotch boiler:* A cylindrical steel shell with one or more cylindrical internal steel furnaces located (generally) in the lower portion and with a bank or banks (passes) of tubes attached to both end closures.

*Sectional header:* A horizontal boiler of the longitudinal or cross drum type, with the tube bank comprised of multiple parallel sections, with each section made up of a front and rear header connected by one or more vertical rows of generating tubes, and with the sections or groups of sections having a common steam drum.

*Vertical:* A fire-tube boiler consisting of a cylindrical shell between the top head and the tube sheet which forms the top of the internal furnace. The products of combustion pass from the furnace directly through the vertical tubes.

Submerged vertical is the same as the plain type above, except that by using a water leg construction as part of the upper tube sheet, it is possible to carry the waterline at a point above the top ends of the tubes.

*Water tube:*  A boiler in which the tubes contain water and steam, with the heat being applied to the outside surface.

*Boiler blowoff piping:*  The piping connections from the boiler to the blow-off valves.

*Boiler blowoff tank:*  A vented and drained container into which water is discharged above atmospheric pressure from the boiler blowoff line.

*Boiler heating surface:*  That part of the heat-transmitting surfaces that are in direct contact with the water or steam on one side, and the fire or heating device on the other side.

*Boiler, high pressure:*  A boiler furnishing steam at pressure in excess of 15 pounds psig or hot water at temperatures in excess of 250°F or at pressures in excess of 160 pounds psig.

*Boiler, high-temperature hot water:*  A water-heating boiler operating at a pressure exceeding 160 psig or temperatures exceeding 250°F.

*Boiler horsepower:*  The evaporation of 34 1/2 pounds of water per hour from a temperature of 212°F into dry saturated steam at the same temperature. Equivalent to 33,472 Btu.

*Boiler, low-pressure hot water and low-pressure steam:*  A boiler furnishing hot water at pressures not exceeding 160 psig or at temperatures not more than 250°F or steam at pressures not more than 15 psig.

*Boiler water:*  A term construed to mean a representative sample of the circulating water, after the generated steam has been separated and before the incoming feed water or added chemical becomes fixed with it so that its composition is affected.

*Boiling out:*  The boiling of a highly alkaline water in boiler pressure parts for the removal of oils, grease, etc. prior to normal operation or after major repairs.

*British thermal unit (Btu):*  The amount of heat required to raise the temperature of one pound of pure water one degree Fahrenheit.

*Bucket trap (inverted):*  A float trap having an open float. The float or bucket is located in the bottom of the trap. The condensate flows into the bucket and when the bucket sinks a valve is opened and the condensate is forced into the return line.

*Bucket trap (open):*  A trap in which the float (bucket) is open at the top. The bucket floats in water which is trapped around it, keeping it floating. The float is attached to a pin which is pressed against its seat. When the bucket fills with condensate, it sinks, pulling the pin from its seat and opening a port through which the condensate is forced into the condensate line.

*Bypass:*  A passage for a fluid, permitting a portion or all of the fluid to flow around certain heat-absorbing surfaces over which it would normally pass.

*Calcium carbonate (CaCO₃):* Usually the chief constituent of boiler scale. Hardness and alkalinity are usually reported as the calcium carbonate equivalent.

*Carbon dioxide:* The most common cause of corrosion in return condensate systems.

*Carryover:* The boiler water solids and liquid entrained with the steam generating unit.

*Cation exchanger:* An ion exchange device which removes hardness from water when regenerated with salt. It removes both hardness and alkalinity when regenerated with acid.

*Chemical feed pipe:* A pipe inside a boiler drum through which chemicals for treating the boiler water are introduced.

*Chimney effect:* The force in a vertical air passage that causes heated air to rise through it. Heated air is less dense than the cooler air.

*Circulation:* The movement of water and steam within a steam generating unit.

*Commercial boiler:* A boiler which produces steam or hot water primarily for heating in commercial applications with incidental use in process applications. Commercial boilers come in a wide range of types, sizes, capacities, pressures, and temperatures. They may also be supplied for more than one application.

*Concentration:* (1) The weight of solids contained in a unit weight of boiler or feed water. (2) The number of times that the dissolved solids have increased from the original amount in the feed water to that in the boiler water due to evaporation in generating steam.

*Condensate:* Condensed water resulting from the removal of latent heat from steam.

*Conductivity:* The property of a water sample to transmit electric current under a set of standard conditions. Usually expressed as micromhos conductance.

*Continuous blowdown:* The uninterrupted removal of concentrated boiler water from a boiler to control total solids concentration in the remaining water.

*Control:* Any manual or automatic device for the regulation of a machine to keep it at normal operation. If automatic, the device is motivated by variations in temperature, pressure, water level, time, light, or other influences.

*Corrosion:* The wasting away of metals due to chemical action in a boiler usually caused by the presence of $O_2$, $CO_2$ or an acid.

*Deaeration:*    The removal of air and gases from boiler feed water prior to its introduction into a boiler.

*Dealkalization:*    An ion exchange process which is used to remove alkalinity from water. This is done with either a cation resin using sulfuric acid generation, or an anion resin using sodium chloride regeneration.

*Degasification:*    The removal of gases from samples of steam taken for purity tests. The removal of $CO_2$ from water in the ion exchange method of softening.

*Demineralizer:*    An ion exchange device used to remove solids from water.

*Design load:*    The load for which a steam generating unit is designed, usually considered the maximum load to be carried.

*Design pressure:*    The pressure used in the design of a boiler for the purpose of determining the minimum permissible thickness or physical characteristics of the different parts of the boiler.

*Design steam temperature:*    The temperature of steam for which a boiler, superheater or reheater is designed.

*Direct-return system (hot water):*    A two-pipe hot water system in which the water passes through the heating device and then returns as directly as possible to the boiler.

*Disengaging surface:*    The surface of the boiler water from which steam is released.

*Dissociation:*    The process by which a chemical compound breaks down into simpler constituents, as $CO_2$ and $H_2O$ at high pressure.

*Dissolved gases:*    Gases which are in solution in water.

*Dissolved solids:*    Those solids in water which are in solution.

*Downfeed one-pipe riser (steam):*    The pipe which carries the steam to the heating units which are located below the steam main. It also carries the condensate back to the boiler.

*Downfeed system (steam):*    A type of steam heating system having the steam main above the heating units it serves.

*Drain:*    A valve connection to the lowest point for the removal of all water from the pressure parts.

*Dry return (steam):*    A return pipe in a steam heating system through which both condensate and air flow back to the boiler.

*Dry steam:*    Steam containing no moisture. Commercially dry steam containing not more than one-half of one percent moisture.

*Economizer:*    A heat recovery device designed to transfer heat from the products of combustion to boiler feed water.

*Efficiency:*    The ratio of output to input. The efficiency of a steam generating unit is the ratio of the heat absorbed by the water and steam to the heat in the fuel used.

*Electric boiler:*   A boiler in which electric heating means serve as the source of heat.

*Entrainment:*   The conveying of particles of water or solids from the boiler water by the steam.

*External treatment:*   Treatment of boiler feed water prior to its introduction into the boiler.

*Feed water:*   Water introduced into a boiler during operation. It includes make-up and return condensate.

*Feed water treatment:*   The treatment of boiler feed water by the addition of chemicals to prevent the formation of scale or corrosion or to eliminate other objectionable characteristics.

*Filming amine:*   An organic chemical which forms a water repellant film when steam condenses. The film controls corrosion in the condensate return.

*Firetube:*   A tube in a boiler having water on the outside and carrying the products of combustion on the inside.

*Flash tank:*   See Blowoff tank.

*Flashing:*   Steam produced by discharging water at saturation temperature into a region of lower pressure.

*Float and thermostatic trap:*   A float that is equipped with a thermostatic element which allows air to pass into the return line.

*Float trap:*   A steam trap which is operated by a float in a chamber. When the float chamber has filled with condensate the float will rise, opening a port through which the condensate flows into the return line. When the condensate level has dropped, the float will close the port to prevent steam from entering the return line.

*Foaming:*   The continuous formation of bubbles which have sufficiently high surface tension to remain as bubbles beyond the disengaging surface.

*Forced circulation:*   The circulation of water in a boiler by a pump, or pumped as opposed to natural circulation.

*Furnace:*   The lower part of a boiler in which the fuel is burned. Also a firebox.

*Gauge cock:*   A valve that is attached to a water column for checking the water level.

*Gauge glass:*   The transparent part of a water gauge assembly connected directly or through a water column to the boiler, below and above the waterline, to indicate the water level in the boiler.

*Gauge pressure:*   The pressure above atmospheric pressure.

*Handhole:*   An opening in a pressure part for access, usually not exceeding 6 inches in the longest dimension.

*Hard water:*   Water which contains calcium or magnesium in an amount which requires an excessive amount of soap to form a lather.

*Hardness:*   A measure of the amount of calcium and magnesium salts in a boiler water. Usually expressed as grains per gallon or ppm as $CaCO_3$.

*Head:*   A unit of pressure which is generally expressed in feet of water.

*Hot water heating system:*   A system which uses the hot water generated by a boiler for heating purposes. The hot water is carried through pipes to the point of heating demand.

*Hydrate alkalinity:*   A measure of caustic or sodium hydroxide in water.

*Industrial boiler:*   A boiler which produces steam or hot water for process applications for industrial use with incidental use for heating. Industrial boilers cover a wide range of sizes, capacities, pressures, and temperatures. They may also be supplied for more than one application (co-generation, etc.).

*Inhibitor:*   A substance which selectively retards a chemical action. An example in boiler work is the use of an inhibitor, when using acid to remove scale, to prevent the acid from attacking the boiler metal.

*Intermittent blowdown:*   The blowing down of boiler water at intervals.

*Internal treatment:*   The treatment of boiler water by introducing chemicals directly into the boiler.

*Ion:*   A charged atom or radical which may be positively or negatively charged.

*Ion exchange:*   A reversible process by which ions are interchanged between solids and a liquid with no substantial structural change of the solid.

*M Alkalinity:*   Sometimes called total alkalinity. Test uses methyl orange indicator.

*Makeup:*   The water added to a boiler feed to compensate for that lost through exhaust, blowdown, leakage, etc.

*Manhole:*   The opening in a pressure vessel of sufficient size to permit a man to enter.

*Moisture in steam:*   Particles of water carried in steam usually expressed as the percentage by weight.

*Natural circulation:*   The circulation of water in a boiler caused by differences in density; also referred to as thermal or thermally induced circulation.

*Neutralizing amine:*   An alkaline organic chemical which neutralizes the acidity of condensate to control corrosion.

*One-pipe supply riser (steam):*   A pipe used to carry steam to an overhead heating unit and to carry the condensate return to the boiler.

*One-pipe system (hot water):*   A hot water heating system which uses one pipe for both the supply and the return. The heating units may have different connections but both are connected to the one pipe.

*One-pipe system (steam):*   A steam heating system which uses only one pipe for both the supply steam and the return condensate. The heating units generally have only one connection which serves as both the supply and the return.

*Overhead system:*   This may be either a steam or a hot water heating system having the supply main located above the heating units. When steam is used the return line must be below the heating units. With hot water the return may be either above or below the units.

*Oxygen attack:*   The corrosion or pitting in a boiler caused by oxygen.

*P Alkalinity:*   A measure of carbonate and hydrate alkalinity using phenolphthalein indicator.

*Packaged boiler:*   See Packaged steam generator.

*Packaged steam generator:*   A boiler which is equipped and shipped complete with fuel-burning equipment, mechanical draft equipment, automatic controls, and accessories. Usually shipped in one or more major sections.

*pH:*   The hydrogen ion concentration of a water to denote acidity or alkalinity. A pH of 7 is neutral. A pH above 7 denotes alkalinity while one below 7 denotes acidity. This pH number is the negative exponent of 10 representing hydrogen ion concentration in grams per liter. For instance, a pH of 7 represents $10^{-7}$ grams per liter.

*Pitting:*   A concentration attack by oxygen or other corrosion chemicals on a boiler, producing a localized depression in the metal surface.

*ppm:*   Abbreviation of parts per million. Used in chemical determinations as one part per million parts by weight.

*Precipitate:*   To separate materials from a solution by the formation of insoluble matter by chemical reaction. The material which is removed.

*Pressure-reducing valve:*   A valve which is used to change the high pressure of a liquid or gas to a lower pressure.

*Pressure vessel:*   A closed vessel container designed to confine a fluid at a pressure above atmospheric.

*Priming:*   The discharge of steam containing excessive quantities of water in suspension from a boiler, due to violent ebullition.

*Process steam:*   The steam used for industrial purposes other than for introducing power.

*Purity:*   The degree to which a substance is free of foreign materials.

*Rate of blowdown:*   A rate normally expressed as a percentage of the incoming water.

*Rated capacity:*   The manufacturer's stated capacity rating for mechanical equipment, for instance, the maximum continuous capacity in pounds of steam per hour for which a boiler is designed.

*Raw water:*   Water supplied to the plant before treatment.

*Residential boiler:*   A boiler which produces low-pressure steam or hot water primarily for heating applications in living quarters or private dwellings.

*Return mains:*   The pipes in a steam or hot water system which are used to return the heating medium from the heating units to the boiler.

*Reverse-return system (hot water):*   A type of hot water system in which the water from the heating units is returned through pipes which are designed so that each heating circuit will have approximately the same length.

*Safety valve:*   A spring-loaded valve that automatically opens when pressure reaches the valve setting. Used to prevent excessive pressure from building up in a boiler.

*Sampling:*   The removal of a portion of a material for examination or analysis.

*Saturated steam:*   Steam at the pressure corresponding to its saturation temperature.

*Saturated water:*   Water at its boiling point.

*Saturation temperature:*   The temperature at which evaporation occurs at a particular pressure.

*Scale:*   A hard coating or layer of chemical material on internal surfaces of boiler pressure parts.

*Scotch boiler:*   See Boiler.

*Secondary treatment:*   The treatment of boiler feed water or the internal treatment of boiler water after primary treatment.

*Sediment:*   Matter in water which can be removed from suspension by gravity or mechanical means.

*Seolite softener:*   See Cation exchanger.

*Shell:*   The outer cylindrical portion of a pressure vessel.

*Sludge:*   A soft water-formed deposit which normally can be removed by blowing down.

*Slug:*   A large "dose" of chemical treatment applied internally to a steam boiler intermittently. Also sometimes instead of "priming" to denote a discharge of water out through a boiler steam outlet in relatively large intermittent amounts.

*Sodium sulfite:*   An oxygen scavenger commonly used in boiler and hot water heating systems.

*Soft water:* Water which contains little or no calcium or magnesium salts, or water from which scale-forming impurities have been removed or reduced.

*Softening:* The act of reducing scale-forming calcium and magnesium impurities from water.

*Solution:* A liquid, such as boiler water, containing dissolved substances.

*Stagnation:* The condition of being free from movement or lacking circulation.

*Steam:* The vapor phase of water substantially unmixed with other gases.

*Steam generating unit:* A unit to which water, fuel and air or wasteheat are supplied and in which steam is generated. It can consist of a boiler, furnace, and fuel-burning equipment, and may include as component parts waterwells, superheater, reheater, air heater, or any combinations thereof.

*Steam purity:* The degree of contamination. Contamination usually expressed in ppm.

*Steam quality:* The percent by weight of vapor in a steam and water mixture.

*Steam separator:* A device for removing the entrained water from the steam.

*Superheat:* To raise the temperature of steam above its saturation temperature. The temperature in excess of its saturation temperature.

*Superheated steam:* Steam at a higher temperature than its saturation temperature.

*Surface blowoff:* The removal of water, foam, etc., from the surface at the water level in a boiler.

*Surge:* The sudden displacement or movement of water in a closed vessel or drum.

*Suspended solids:* Undissolved solids in boiler water.

*Sweat:* The condensation of moisture from a warm saturated atmosphere on a cooler surface. A slight weep in a boiler joint but not in a sufficient amount to form drops.

*Swell:* The sudden increase in the volume of the steam in the water-steam mixture below the water level.

*Total solids in concentration:* The weight of dissolved and suspended impurities in a unit weight of boiler water, usually expressed in ppm.

*Treated water:* The water which has been chemically treated to make it suitable for boiler feed.

*Turbity:* The optical obstruction to the passing of a ray of light through a body of water, caused by finely divided, suspended matter.

*Vapor:* The gaseous product of evaporation.

*Vent:*   An opening in a vessel or other enclosed space for the removal of gas or vapor.

*Vent valve:*   A valve which is designed to permit air to escape from either a steam or a hot water system. They are generally located in the highest point in the system or where air entrapment may be a problem.

*Vent valve, main:*   A vent valve which is located in the main supply line at the furthest point from the boiler. It is used to vent air from the mains so that the heating medium can start circulating more quickly.

*Water column:*   A vertical tubular member connected at its top and bottom, respectively, to the steam and water space of a boiler, to which the water gauge, gauge cocks, and the high and low water level alarms may be connected.

*Water gauge:*   The gauge glass and its fittings for attachment.

*Water leg:*   A vertical or nearly vertical box header or sectional header, or water-cooled sides of an internal firebox composed of flat or circular surfaces.

*Water level:*   The elevation of the surface of the water in a boiler.

*Water tube:*   A tube in a boiler having the water and steam inside and the heat applied to the outside.

*Water vapor:*   A synonym for steam, usually used to denote steam of low absolute pressure.

*Weep:*   A term usually applied to a minute leak in a boiler joint which forms droplets (or tears) of water very slowly.

*Wetness:*   A term used to designate the percentage of water in steam. Also used to describe the presence of a water film on heating surface interiors.

*Wet steam:*   Steam containing moisture.

# APPENDIX

| To convert | Into | Multiply by |
|---|---|---|
| Calories | BTUs | 0.003968 |
| Calories | Joules | 4.186 |
| Centimeters | Inches | 0.3937 |
| Degrees Celsius | Degrees Fahrenheit | (Degrees Celsius $\times$ 1.8) + 32 |
| Degrees Fahrenheit | Degrees Celsius | (Degrees Fahrenheit − 32) (5/9) |
| Feet | Centimeters | 30.48 |
| Feet | Meters | 0.3048 |
| Grams | Pounds | 0.00220 |
| Inches | Centimeters | 2.54 |
| Inches | Millimeters | 25.40 |
| Joules | BTUs | 0.0009478 |
| Joules | Calories | 0.2389 |
| Kilocalories | Calories | 1000 |
| Kilograms | Pounds | 2.205 |
| Kilopascals | Pounds per square inch | 0.1450326 |
| Liters | Gallons (Imperial) | 0.220 |
| Liters | Gallons (U.S. liquid) | 0.264 |
| Meters | Feet | 3.281 |
| Meters | Inches | 39.37 |
| Millimeters | Inches | 0.03937 |
| Pounds | Grams | 453.6 |
| Pounds per square inch | Kilopascals | 6.895 |

# INDEX